The Logica Yearbook
2018

The Logica Yearbook 2018

Edited by

Igor Sedlár
and
Martin Blicha

© Individual authors and College Publications 2019
All rights reserved.

ISBN 978-1-84890-307-4

College Publications
Scientific Director: Dov Gabbay
Managing Director: Jane Spurr

www.collegepublications.co.uk

Original cover design by Laraine Welch

All rights reserved. No part of this publication may be reproduced, stored in a retrieval system or transmitted in any form, or by any means, electronic, mechanical, photocopying, recording or otherwise without prior permission, in writing, from the publisher.

Preface

This volume contains a selection of the contributions presented at the symposium *Logica 2018*, which took place at the Hejnice Monastery in northern Czech Republic between June 18 and June 22, 2018.

The tradition of Logica symposia is a long one, starting in 1987. The symposium is open to researchers of both a mathematical and a philosophical bent, and one of it's main aims is to foster a fruitful exchange of ideas between these overlapping groups. The informal atmosphere aims to provide space for a stimulating dialogue between logicians of all generations, including students. As the editors of this volume, we are proud that we can contribute to the successful completion of the annual symposium cycle by presenting this collection to you.

Logica 2018 was, just as all the previous Logica symposia, organized by the Department of Logic of the Institute of Philosophy of the Czech Academy of Sciences. More than thirty lectures were presented during the conference, including those given by our distinguished invited speakers Samson Abramsky, Francesco Berto, Danielle Macbeth and Jaroslav Peregrin. A tutorial was given by Carles Noguera. Unfortunately, the proceedings, traditionally published within one year of the conference, can offer only a limited record of the topics discussed and cannot hope to even partially convey the symposium atmosphere. Nevertheless, we believe that the articles in this volume can more than stand on their own.

Both the Logica symposium and The Logica Yearbook are the result of a joint effort of many people to whom we would like to express our gratitude. We are, of course, very grateful to the Institute of Philosophy of the Czech Academy of Sciences for all the support that made the event possible. We would also like to thank the staff of Hejnice Monastery for their hospitality and friendly assistance. We gratefully acknowledge the support of the Bernard Family Brewery of Humpolec, the traditional sponsor of the Logica social programme. We owe thanks also to the Czech Science Foundation, which provided significant support for the meeting and for the publication of this book with funding from the grant project no. 17-15645S. Ondrej Majer designed the trademark Logica T-shirts. Olga Bažantová proved again to be a key member of the organizing team by dealing with administrative and practical matters efficiently and with wit. We are of course grateful also to College Publications and its managing director, Jane Spurr, for our consistently pleasant cooperation during the preparation of each Logica Yearbook

Series volume. Last but not least, we would like to thank all of the authors for their contributions and collaboration during the editorial process.

Prague, May 2019

Igor Sedlár and Martin Blicha

Table of Contents

Defining Selection Functions 1
    Holger Andreas and Mario Günther

Is Logic Exceptional? 13
    Pavel Arazim

The Theory of Topic-Sensitive Intentional Modals 31
    Francesco Berto

Categoricity and Negation. A Note on Kripke's Affirmativism 57
    Constantin C. Brîncuş and Iulian D. Toader

A Justification Logic for Argument Evaluation 67
    Alfredo Burrieza and Antonio Yuste-Ginel

On the Nature of Combinatorial Proofs 83
    Serena Delli

A Formal Model for Explicit Knowledge as Awareness-Of Plus
Awareness-That 101
    Claudia Fernández-Fernández and Fernando R. Velázquez-Quesada

Faithful Relative Truth Definability and a Truth-Theoretic Strength
of Type-Free Truth 117
    Daichi Hayashi

Toward a Logic for Neural Networks 133
    Levin Hornischer

The Name of the Sinus Function 149
    Ansten Klev

Algebraic Semantics for Fischer Servi's Version of Modal Logic **BK** .. 161
    Sergei Odintsov

Logic as a (Natural) Science 177
    Jaroslav Peregrin

Non-Constructive Procedural Theory of Propositional Problems and
the Equivalence of Solutions 197
    Ivo Pezlar

A First-Order Sequent Calculus for Logical Inferentialists and Expressivists..211
   Shuhei Shimamura

A Dynamic Epistemic Logic for Resource-Bounded Agents 229
   Anthia Solaki

Defining Selection Functions

HOLGER ANDREAS AND MARIO GÜNTHER

Abstract: In AGM-style belief revision theory, various types of epistemic ordering have been described. First, we argue that the notion of an epistemic ordering on a finite belief base is cognitively most plausible and most suitable for the analysis of specific examples. Second, we show how other types of epistemic ordering can be derived from an epistemic ordering on a finite belief base.

Keywords: belief revision, epistemic ordering, partial meet contraction

1 Introduction

The notion of an epistemic ordering is essential to AGM belief revision theory, founded by Alchourrón, Gärdenfors, and Makinson (1985). Various types of such an ordering have been described by the authors of the AGM theory and others working in the AGM framework. Each type gives rise to at least one distinct belief revision scheme. In essence, we can distinguish between the following types of epistemic ordering:

1. ordering on a finite belief base H

2. ordering on the union of the remainder sets of a finite belief base H

3. ordering on a logically closed belief set K

4. ordering on the union of the remainder sets of a logically closed belief set K

5. plausibility ordering on possible worlds.

Arguably, the notion of an epistemic ordering on a finite belief base H is cognitively the most plausible type of ordering. When we come to analyse concrete examples of belief changes, it is by far easier to specify an epistemic ordering on a finite belief base than on an infinite belief set. Directly specifying an epistemic ordering on a logically closed belief set is, strictly speaking, beyond the cognitive capacities of a finitely bounded human mind since such a set is an infinite entity. Hence, it is desirable to have a means for determining an epistemic ordering on a logically closed belief set using an epistemic ordering on a finite belief base.

Similar considerations apply to the problem of directly specifying an epistemic ordering on the union of the remainder sets of a finite belief base. (The remainder set $H \perp \alpha$ is the set of a maximal subsets of H that do not entail α; see Section 3 for the precise definition.) Even though the members of such a set are finite entities, it is difficult to form intuitions about how entrenched a finite set of beliefs is when taken as a whole unit. While this is not a problem for proving representation theorems about belief changes, it turns out to be a problem when we come to work out concrete applications of belief revision theory, including the analysis of conditionals, causal relations, and scientific explanations. In this context, recall that a cognitively more plausible analysis of conditionals was a key motivation for Gärdenfors to study belief changes.[1] Hence, considerations of cognitive plausibility and applicability should matter for our theories of belief revision.

Let us, finally, discuss epistemic orderings of type (4) and (5). If directly specifying an epistemic ordering on a logically closed belief set is beyond the cognitive means of a finitely bounded human mind, then the direct specification of an epistemic ordering on the union of remainder sets of a logically closed belief set is a fortiori so. Possible worlds furnished with a plausibility ordering have great intuitive appeal. At the same time, we think that the plausibility of a world depends on which sentences are true in this world, but not the other way around. We do not have intuitions about the plausibility of a possible world taken as a whole unit, but about sentences that hold true in this world. For example, we say that possible worlds in which presently presumed laws of physics hold true are more plausible than others in which some of these laws do not hold. Notice, moreover, that Lewis (1979) specifies some guidelines about the similarity ordering of possible worlds in terms of differences among possible worlds at the level of singular facts and laws.

In sum, we can say that the notion of an epistemic ordering on a finite belief base has a relatively well defined, direct empirical content, in the sense that some of our explicit beliefs are more firmly established than others. Other types of epistemic ordering seem to be more theoretical, in the sense of being less directly related to our explicit beliefs. It is therefore

---

[1] See the foreword by David Makinson of the 2008 reprint of (Gärdenfors, 1988) by College Publications, edited by Dov Gabbay.

desirable to investigate if we can derive, respectively, an epistemic ordering of the types (2) to (5) from an epistemic ordering on a finite belief base, thereby relating certain theoretical concepts of belief revision theory to concepts with a more direct empirical content. This is the problem to which the present paper is addressed.

Another motivation for pursuing the present investigation is that belief revision schemes with an epistemic ordering on a finite belief base have received relatively little attention in the belief revision literature. As observed by Hansson (1999, p. 103), it proved difficult to define entrenchment-based revisions for belief bases in a manner closely analogous to entrenchment-based belief set revisions. The theory of ensconcement relations by Williams (1994) is, arguably, too radical a proposal because the corresponding revision operation "cuts away" unnecessarily many beliefs. The preferred subtheory approach by Brewka (1991) is simple and intuitive, but does not satisfy all postulates of belief base revision established by Hansson (1993), and so deviates more strongly from the original AGM theory than Hansson's seminal account of belief base revision (Hansson, 1993). The latter account, however, works with selection functions, which in turn are based on an epistemic ordering on the union of the remainder sets of a belief base.

The overall strategy for deriving epistemic orderings of type (2) to (4) from an epistemic ordering of type (1) is to define selection functions of appropriate type on the basis of a strict weak ordering on a finite belief base. This directly gives us an epistemic ordering of type (2) and (4). An epistemic entrenchment ordering on a logically closed belief set can be defined via partial meet contractions, which in turn are based on selection functions. A plausibility ordering on a set of possible worlds will be defined in terms of degrees of satisfaction of a finite belief base with priorities.

2   Levels of epistemic priority

Let H be a finite belief base. That is, H is a set of sentences that are intended to represent explicit beliefs. Drawing on the work of Brewka (1991), we assume the epistemic ordering among the items of H to be a strict weak ordering. Such an ordering can be represented by a sequence of subsets of H:

$$\mathbf{H} = \langle H_1, \ldots, H_n \rangle.$$

where $H_1, \ldots, H_n$ is a partition of H. **H** is called a *a prioritised belief base*. The indices represent an epistemic ranking of the beliefs. H_1 is the set of the most firmly established beliefs, the beliefs in H_2 have secondary priority, etc. In other words, partitioning H into disjoint subsets is intended to introduce different levels of epistemic priority into the "flat" belief base H.

It must be acknowledged that the assumption of a strict weak ordering on H is still an idealization. For some pairs of explicit beliefs, we may lack an intuition as to which belief is more firmly established, and it may not always be appropriate to infer from this that the two beliefs are equally firm. Partial orderings of a belief base are considered by Brewka (1991), but in this paper we shall not address the problem of deriving epistemic orderings on the basis of a partially ordered belief base.

3 Partial meet revision

Hansson (1993) defines (internal) partial meet base revisions in a manner closely analogous to partial meet revisions of Alchourrón et al. (1985). Both belief revision schemes hinge on the notion of a selection function, defined for remainder sets of the belief base and the belief set, respectively. Drawing on the presentation of Hansson (1999, Ch. 1), we recall the core definitions underlying the two belief revision schemes. Let us begin with the notion of a *remainder set* of A relative to α, where A may or may not be finite.

Definition 1 $A \perp \alpha$
Let A be a set of sentences and α a sentence. $A' \in A \perp \alpha$ iff

1. $A' \subseteq A$

2. $\alpha \notin Cn(A')$

3. there is no A'' such that $A' \subset A'' \subseteq A$ and $\alpha \notin Cn(A'')$.

$Cn(A)$ designates the logical closure of A. Some members of a remainder set $A \perp \alpha$ may be epistemically superior to others, in the sense that their beliefs are more firmly held than the beliefs of other members of $A \perp \alpha$. At this point, the notion of a selection function comes into play. Suppose $\leq$ is a binary transitive relation on the union of all remainder sets $A \perp \alpha$. $B \leq B'$ means that B' is epistemically not inferior to B. Equivalently, $B \leq B'$ means that B' is epistemically at least as good as B. Using such an epistemic ordering, we can define a selection function for the remainder set as follows:

$$\sigma(A\perp\alpha) = \{A' \in A\perp\alpha \mid A'' \leq A' \text{ for all } A'' \in A\perp\alpha\}. \quad \text{(Def } \sigma\text{)}$$

Then, we take the selected members of the remainder set to define the contraction of A by α:

$$A - \alpha = \bigcap \sigma(A\perp\alpha). \quad \text{(PMC)}$$

Now we are in a position to define partial meet revisions using the Levi identity $(A * \alpha = (A - \alpha) + \alpha)$:

$$A * \alpha = \bigcap \sigma(A\perp\neg\alpha) + \alpha. \quad \text{(PMR)}$$

In line with the presentation of Hansson (1999, Ch. 1), we have not made any assumption as to whether the set A of sentences is finite or not. Hence, A may be a finite belief base or a logically closed belief set. Note, however, that belief bases and belief sets differ as regards the definition of expansions, represented by the symbol $+$.

4 Defining selection functions

How can we determine a selection function based on some intuition about an epistemic ranking of sentences in $H \subseteq A$? The rationale of the following definitions is to select those members of $A\perp\alpha$ that overlap with the more firmly established beliefs in H to the greatest extent. Let us make this idea more precise:

Definition 2 *Let A be a set of sentences, and A'' and A' be subsets of A. $A'' < A'$ iff there is i $(1 \leq i \leq n)$ such that*

1. $A'' \cap H_i \subset A' \cap H_i$, and

2. for all $j < i$ $(j \geq 1)$, $A'' \cap H_j = A' \cap H_j$.

Definition 3 $A' \equiv A''$ iff $A' \cap H_i = A'' \cap H_i$ for all i $(1 \leq i \leq n)$.

Definition 4 $A' \leq A''$ iff $A' < A''$ or $A' \equiv A''$.

Since $\leq$ is defined on $A \times A$, it is well defined for the union of all remainder sets of A.

In the classical AGM theory, it is desirable that the relation $\leq$ underlying the selection function σ is transitive. If this condition is satisfied, it can be shown that the revision function $*$ – defined by (Def σ), (PMC), and (PMR) – satisfies the postulates $K*1$ to $K*8$ (see Theorems 3.1 and 4.16 in Gärdenfors, 1988). As is well known, these postulates are considered canonical for belief set revisions in the AGM theory. For this reason, let us show that the relation $\leq$ defined by Definition 4 is in fact transitive, using the transitivity of the relation $<$ defined by Definition 2.

Proposition 1 *If $A''' < A''$ and $A'' < A'$, then $A''' < A'$.*

Proof. Suppose (i) $A''' < A''$ and (ii) $A'' < A'$. By Definition 2, there are i and l ($1 \leq i, l \leq n$) such that

(i) $A''' \cap H_i \subset A'' \cap H_i$ and for all $j < i$ ($j \geq 1$), $A''' \cap H_j = A'' \cap H_j$, and

(ii) $A'' \cap H_l \subset A' \cap H_l$ and for all $k < l$ ($k \geq 1$), $A'' \cap H_k = A' \cap H_k$.

We consider now three exhaustive cases:

(a) $l = i$: Then, there is some i such that $A''' \cap H_i \subset A'' \cap H_i = A'' \cap H_l \subset A' \cap H_l$ and $A''' \cap H_k = A'' \cap H_k = A' \cap H_k$ for all $k < l$ ($k \geq 1$). Because of the transitivity of the strict subset relation $\subset$ and Definition 2, it holds that $A''' < A'$, as desired.

(b) $l > i$: Then, there is some i such that $A''' \cap H_i \subset A'' \cap H_i = A' \cap H_i$ and $A''' \cap H_j = A'' \cap H_j = A' \cap H_j$ for all $j < i$ ($j \geq 1$). By the transitivity of the strict subset relation $\subset$ and Definition 2, we can infer from this that $A''' < A'$, as desired.

(c) $l < i$: Then, there is some l such that $A''' \cap H_l = A'' \cap H_l$, but $A'' \cap H_l \subset A' \cap H_l$. Hence, $A''' \cap H_l \subset A' \cap H_l$. From $l < i$ we can furthermore infer that $A''' \cap H_k = A'' \cap H_k = A' \cap H_k$ for all $k < l$. By Definition 2, these two conclusions imply that $A''' < A'$, as desired.

Thus, we have shown that $<$ defined by Definition 2 is transitive. □

We show now that $\leq$ defined by Definition 4 is transitive as well.

Proposition 2 *If $A''' \leq A''$ and $A'' \leq A'$, then $A''' \leq A'$.*

Proof. Assume $A''' \leq A''$ and $A'' \leq A'$. By Definition 4, this implies $A''' < A''$ or $A''' \equiv A''$, and $A'' < A'$ or $A'' \equiv A'$. We need to consider four cases.

Defining Selection Functions

(a) Suppose $A''' < A''$ and $A'' < A'$. Then, by Proposition 1, $A''' < A'$. By Definition 4, $A''' \leq A'$.

(b) Suppose $A''' < A''$ and $A'' \equiv A'$. Then, by Definition 2, there is i $(1 \leq i \leq n)$ such that (i) $A''' \cap H_i \subset A'' \cap H_i$ and (ii) for all $j < i$ $(j \geq 1)$, $A''' \cap H_j = A'' \cap H_j$. By Definition 3, $A'' \equiv A'$ implies that, for all k $(1 \leq k \leq n)$, $A'' \cap H_k = A' \cap H_k$. Using (i) and (ii), we can infer from this that (iii) $A''' \cap H_i \subset A' \cap H_i$ and (iv) for all $j < i$ $(j \geq 1)$, $A''' \cap H_j = A' \cap H_j$. By Definition 4, $A''' \leq A'$.

(c) Case (c) is identical to case (b), if we simply switch $<$ and $\equiv$ such that the supposition is $A''' \equiv A''$ and $A'' < A'$.

(d) Suppose $A''' \equiv A''$ and $A'' \equiv A'$. Then, by Definition 3, for all i $(1 \leq i \leq n)$, $A''' \cap H_i = A'' \cap H_i$ and $A'' \cap H_i = A' \cap H_i$. Hence, by the transitivity of equality $=$ on sets, $A''' \cap H_i = A' \cap H_i$ for all i $(1 \leq i \leq n)$. By Definition 4, we obtain $A''' \equiv A'$, and thus $A''' \leq A'$, as desired.

Thus, we have shown that $\leq$ defined by Definition 4 is transitive. □

5 Epistemic entrenchment

Entrenchment-based revisions are at the core of the classical AGM theory. Let us begin with the formal characterisation of the epistemic entrenchment relation. $\alpha \leq_e \beta$ means that α is at most as entrenched as β. The following postulates formally characterise this relation (Gärdenfors, 1988, Ch. 4.6):

If $\alpha \leq_e \beta$ and $\beta \leq_e \chi$, then $\alpha \leq_e \chi$	(EE1)
If $\alpha \vdash \beta$, then $\alpha \leq_e \beta$	(EE2)
$\alpha \leq_e \alpha \wedge \beta$ or $\beta \leq_e \alpha \wedge \beta$	(EE3)
When $K \neq K_\bot, \alpha \notin K$ iff $\alpha \leq_e \beta$ for all $\beta \in K$	(EE4)
If $\beta \leq_e \alpha$ for all $\beta \in \mathcal{L}$, then $\alpha \in Cn(\emptyset)$.	(EE5)

where $\mathcal{L}$ is the set of all formulas of the formal language used to analyse belief changes, and $K_\bot$ the absurd belief set containing all elements of $\mathcal{L}$.

As shown by Gärdenfors and Makinson (1988), epistemic entrenchment orderings and contractions are interdefinable by the following equations:

$$\begin{array}{c} \beta \in K - \alpha \text{ iff } \beta \in K \\ \text{and either } \alpha <_e (\alpha \vee \beta) \text{ or } \alpha \in Cn(\emptyset). \end{array} \quad \text{(G-)}$$

$$\beta \leq_e \alpha \text{ iff } \beta \notin K - (\alpha \wedge \beta) \text{ or } \alpha \wedge \beta \in Cn(\emptyset). \qquad (C\leq)$$

The idea underlying (C≤) is that a contraction of K with $\alpha \wedge \beta$ results in the retraction of α or β (or both), and β should be retracted iff α is at least as epistemically entrenched as β. Note that, according to (C≤), $\beta < \alpha$ means (in the non-tautological case) that $\alpha \in K - (\alpha \wedge \beta)$. It goes without saying that $<_e$ is the strict part of the relation $\leq_e$.

Note that an ordering of epistemic entrenchment must satisfy certain logical constraints that may not hold for the epistemic ranking of a prioritised belief base **H**. For this reason, we cannot simply extend the epistemic ranking of a prioritised belief base **H** in order to obtain an epistemic entrenchment ordering. But we can determine an epistemic entrenchment ordering using (C≤) and partial meet contractions defined on the basis of a selection function σ, which in turn is determined by a prioritised belief base **H** via Definition 4. To be more precise:

Proposition 3 *Let $-$ be a contraction function defined on the basis of a prioritised belief base* **H**, *via definitions (Def σ), (PMC) and 4. Further, let $\leq_e$ be defined using this contraction function via (C≤). $\leq_e$ thus defined satisfies the postulates $(EE1) - (EE5)$ of a standard relation of epistemic entrenchment.*

This proposition follows from Theorems 4.16 and 4.30 in (Gärdenfors, 1988).

6 Plausibility ordering of possible worlds

The epistemic orderings considered so far are syntactic in nature since they are defined on sets or sets of sets of sentences. However, an epistemic ordering can also be represented by a system of spheres of possible worlds, as shown by Grove (1988) and Spohn (1988). The basic ideas of this representation are as follows. The possible worlds of the innermost sphere are considered most plausible in the epistemic state to be modelled. The further away a possible world is from the innermost sphere, the less plausible this world is considered.

To describe the formal properties of such a system of spheres, a few more symbols need to be introduced. If α is a sentence, $[\alpha]$ designates the set of possible worlds that verify α. If A is a set of sentences, let $[A]$ designate the set of possible worlds that verify all members of A. $Sent(w)$, by contrast, is the set of sentences that are verified by the possible world w. Now we are

Defining Selection Functions

in a position to define the notion of a system of spheres along the lines of Grove (1988):

Definition 5 *Let W be a set of possible worlds, $\mathcal{L}$ a propositional language, and let K be a belief set. Further, let $<_p$ be a binary relation on W. $\langle W, <_p \rangle$ is a system of spheres centered on K iff*

(S1) $\leq$ is a strict, weak order on W, i.e., asymmetric, transitive, and modular

(S2) $[K] = min_{<_p}(W)$

(S3) $[T] = W$, where T is a tautology

(S4) $min_{<_p}(W \cap [\alpha])$ is well defined for all sentences α of $\mathcal{L}$.

$min_{<_p}(W')$ designates the set of $<_p$-minimal elements in a set $W' \subseteq W$ of possible worlds. In more formal terms:

$$min_{<_p}(W') = \{w \mid \text{there is no } w' \text{ such that } w' <_p w\}$$

$min_{<_p}(W \cap [\alpha])$ is the set of possible worlds considered most plausible after a revision by α. This suggests the following definition of revisions:

$$K * \alpha = \{\beta \mid w \models \beta \text{ for all } w \in min_{<_p}(W \cap [\alpha])\}.$$

Drawing on Benferhat, Dubois, Lang, and Prade (1993), we can define a system of spheres on the basis of a prioritised belief base:

Definition 6 *Let $\mathbf{H}$ be a finite prioritised belief base, and $\mathcal{L}$ a propositional language in which the sentences of $\mathbf{H}$ are given. Let W be a set of possible worlds such that $[T] = W$. $w <_p w'$ iff there is i $(1 \leq i \leq n)$ such that*

1. $|Sent(w') \cap H_i| < |Sent(w) \cap H_i|$, and

2. for all $j < i$ $(j \geq 1)$, $|Sent(w') \cap H_j| = |Sent(w) \cap H_j|$.

The idea underlying this definition is that a possible world w is more plausible than another w' iff w satisfies more sentences at a certain epistemic level i of $\mathbf{H}$ (in the sense of satisfying a set $H'_i \subseteq H_i$ of sentences with a greater cardinality), while being on a par with w' at the epistemic levels $j < i$.

Proposition 4 *The relation $<_p$ (defined by Definition 6) is asymmetric, transitive, and modular.*

Proof. (1) Asymmetry. The proof is trivial and can be obtained using that the relation $<$ on cardinal numbers is asymmetric.

(2) Transitivity. The proof is analogous to that of Proposition 1, and can be obtained using the fact that the relation $<$ on cardinal numbers is transitive.

(3) Modularity. Modularity is also referred to as the property of being negatively transitive: if, $w' \not< w''$, $w'' \not< w'$, $w'' \not< w'''$, and $w''' \not< w''$, then $w' \not< w'''$ and $w''' \not< w'$. That is, the relation of being incomparable (according to $<_p$) is transitive. Suppose (i) $w' \not< w''$ and $w'' \not< w'$, and (ii) $w'' \not< w'''$ and $w''' \not< w''$. (i) implies (iii) that, at any level j ($1 \leq j \leq n$), $|Sent(w') \cap H_j| = |Sent(w'') \cap H_j|$. (ii) implies (iv) that, at any level j ($1 \leq j \leq n$), $|Sent(w'') \cap H_j| = |Sent(w''') \cap H_j|$. (iii) and (iv) imply that, at any level j ($1 \leq j \leq n$), $|Sent(w') \cap H_j| = |Sent(w''') \cap H_j|$. Hence, $w' \not< w'''$ and $w''' \not< w'$. □

Proposition 5 *The relation $<_p$ (defined by Definition 6) has a well-defined minimum on any set $W' \subseteq W$.*

Proof. Suppose, for contradiction, that $<_p$ has not a well defined minimum on $W' \subseteq W$. By transitivity of $<_p$, this implies that there is an infinite sequence $w_i, w_{i+1}, w_{i+2}, \ldots$ such that $w_i < w_j$ if $i > j$. Hence, for any world w of this infinite sequence, there is another world w' that satisfies, at some epistemic level k, more sentences of H_k than w (in the sense of satisfying a set of sentences with a greater cardinality), while being on a par with w at the levels $l < k$. This implies that **H** has at least one component that is an infinite set. Contradiction. □

From propositions 4 and 5, we can infer that $<_p$ defines a system of spheres of possible worlds (in the sense of Definition 5), which is centred on $[K] = min_{<_p}(W)$.

Let us briefly indicate why we cannot define a system of spheres of possible worlds using a definition of $<_p$ that is closely analogous to Definition 2 (which defines an ordering on sets of sentences). Suppose we define $w <_p w'$ iff there is i ($1 \leq i \leq n$) such that

1. $Sent(w') \cap H_i \subset Sent(w) \cap H_i$, and

2. for all $j < i$ ($j \geq 1$), $Sent(w') \cap H_j = Sent(w) \cap H_j$.

The order thus defined is not always modular, and so qualifies only as a strict partial order. Here is a simple counterexample to modularity. $\mathbf{H} = \langle\{s,p,q,r\}\rangle$. Let us represent possible worlds w by sets of literals such that, for any literal l, l or its complement is a member of w. $w' = \{p,r,\neg q,\neg s\}$, $w'' = \{p,q,\neg r,\neg s\}$, and $w''' = \{p,s,r,\neg q\}$. Then, we have $w' \not< w''$, $w'' \not< w'$, $w'' \not< w'''$, and $w''' \not< w''$, but $w''' < w'$.

7 Conclusion

AGM-style belief revision theory comes with a variety of epistemic orderings, each of which gives rise to at least one specific belief revision scheme. Of these types of epistemic ordering, the notion of an epistemic ordering on a finite belief base is most closely and most directly related to our intuitions about some beliefs being more firmly established than others. Other types of epistemic ordering are comparable to theoretical concepts in scientific theories in at least two respects. First, they are less directly related to our intuitions about some beliefs being more entrenched than others. Second, they are motivated by elements of the respective belief revision scheme to a greater extent than the notion of an epistemic ordering on a finite belief base is. For example, the need for an epistemic ordering on a logically closed belief set arises with $K*1$, which says that belief sets are logically closed. Hence, it is desirable to investigate if the more theoretical epistemic orderings can be determined on the basis of an epistemic ordering on a finite belief base. We have answered this question in the affirmative. To be more precise, we have shown how an epistemic ordering on unions of remainder sets, an epistemic entrenchment and a plausibility ordering can be determined on the basis of the epistemic ranking of a finite belief base.

References

Alchourrón, M. A., Gärdenfors, P., & Makinson, D. (1985). On the logic of theory change: Partial meet contraction functions and their associated revision functions. *Journal of Symbolic Logic, 50*, 510–530.

Benferhat, S., Dubois, D., Lang, J., & Prade, H. (1993). Inconsistency management and prioritised syntax-based entailment. In *Proceedings of the 14th International Joint Conference on Artificial Intelligence (IJCAI-93)* (pp. 640–645). Los Altos: Morgan Kaufmann.

Brewka, G. (1991). Belief revision in a framework for default reasoning. In *Proceedings of the Workshop on The Logic of Theory Change* (pp. 602–622). London: Springer.

Gärdenfors, P. (1988). *Knowledge in Flux.* Cambridge: MIT Press.

Gärdenfors, P., & Makinson, D. (1988). Revision of knowledge systems using epistemic entrenchment. In M. Vardi (Ed.), *TARK' 88 - Proceedings of the Second Conference on Theoretical Aspects of Reasoning about Knowledge* (pp. 83–95). Los Altos: Morgan and Kaufmann.

Grove, A. (1988). Two modellings for theory change. *Journal of Philosophical Logic, 17*(2), 157–170.

Hansson, S. O. (1993). Reversing the Levi identity. *Journal of Philosophical Logic, 22*(6), 637–669.

Hansson, S. O. (1999). *A Textbook of Belief Dynamics. Theory Change and Database Updating.* Dordrecht: Kluwer.

Lewis, D. (1979). Counterfactual dependence and time's arrow. *Noûs, 13*(4), 455–476.

Spohn, W. (1988). Ordinal conditional functions: A dynamic theory of epistemic states. In *Causation in Decision, Belief Change, and Statistics II* (pp. 105–134). Dordrecht: Kluwer.

Williams, M.-A. (1994). On the logic of theory base change. In C. MacNish, D. Pearce, & L. M. Pereira (Eds.), *Logics in Artificial Intelligence* (pp. 86–105). Berlin, Heidelberg: Springer.

Holger Andreas
University of British Columbia
Canada
E-mail: `holger.andreas@ubc.ca`

Mario Günther
Universität Regensburg
Germany
E-mail: `mario.guenther@ur.de`

Is Logic Exceptional?

Pavel Arazim[1]

Abstract: Some authors such as Ole Hjorland have recently propounded the thesis that logic is not a particularly exceptional discipline when it comes to how we revise its theses. This view is understood as a reaction to a more traditional view that logic in some important sense is exceptional or even completely unrevisable. I question the very notion of holding a given logical law to be valid and reasoning accordingly. I illustrate problems of both exceptionalism and anti-exceptionalism, as both the opposite theses appear in Quine. Possibilities of steering clear from both the radical opposite theses are hinted at.

Keywords: exceptionalism, holism, meaning, logical expressions, Quine

Is logic in some important sense exceptional? This question has become a topic of discussion recently, although it has deeper historical roots. What can we mean when we ask whether logic is exceptional? Obviously, it has to stand out in some way, in order to be distinguishable from other disciplines. That it stands out in this way is not as trivial as it might seem to be, because in case of any purported discipline we can ask whether it is self-standing and thus a genuine discipline. This can be falsified in two opposite ways, as it can either be shown that the purported discipline can be thought of only as a part of a broader and genuinely self-contained discipline or it can be shown that the purported discipline is just an aggregate of disciplines which do not naturally interrelate and are thus better kept separate. The first option was argued for in case of arithmetic by Frege who introduced logicism, i.e., the thesis that arithmetic is not a self-standing discipline, it is rather a part of logic. Some successors tried to show that the whole of mathematics should be in this way embedded in logic.

More recently, the opposite proposals to consider rather logic as part of mathematics have been formulated by many authors. In particular Gila Sher, a leading figure among the model-theoretic demarcators of logical constants (see Sher, 1991). Furthermore, various logics, starting with the classical

---

[1] This article was supported by grant 17-15645S *Logical models of reasoning and argumentation in natural language*, led by Jaroslav Peregrin from Czech Academy of Sciences.

propositional logic, can be studied as specific kinds of algebras. It is thus far from clear that logic is a self-standing discipline.

It will be another kind of exceptionality which we will focus on, though. This further kind of exceptionality concerns the way specifically logical claims are verified and possibly falsified. The epistemological status of logic is thus at stake. Logical exceptionalism claims that logic is exceptional in the sense that it cannot be subject to revision in the way other disciplines are. Ole Hjortland (2017) in his article *Anti-Exceptionalism about logic* sees exceptionalism as a view which was so far at least implicitly accepted and offers his *Anti-Exceptionalism* as an antidote. We will follow what the names suggest, namely that exceptionalism is prior, and discuss it before the anti-exceptionalism. We will illustrate the dilemma about exceptionalism and its opposite by discussing how it is present in Quine's philosophy. I believe revisiting Quine is not only of historical interest, as he manages to motivate both the opposite stances very clearly. Furthermore, I believe studying Quine helps us to see that both exceptionalism and anti-exceptionalist do not really bring so much new into philosophy of logic.

1 Exceptionalism

The core idea of exceptionalism is that logic has a special epistemic status, specifically logical claims are thus verified or falsified in a substantially different manner than in, say, physics. In its most radical form, the thesis could be put thus:

(LE) Logic cannot be revised.

As radical as the thesis appears to be in face of the sheer multitude of logical systems that have been developed so far, it is very natural to think it must be true. And indeed, perhaps ignoring some nuances, something like this thesis has been accepted during the long history of logic more or less unanimously, at least until well into twentieth century when more logical systems such as classical logic or intuitionistic logic and others established themselves. During many periods of the history, this conviction might have remained rather implicit because it seemed so self-evident. Yet it also received straightforward formulations as the famous one due to Kant[2].

---

[2]Daß die Logik diesen sicheren Gang schon von den ältesten Zeiten her gegangen sei, läßt sich daraus ersehen, daß sie seit dem Aristoteles keinen Schritt rückwärts hat tun dürfen, wenn man ihr nicht etwa die Wegschaffung einiger entbehrlicher Subtilitäten, oder deutlichere Bes-

Is Logic Exceptional?

Kant says that logic has been essentially completed already by Aristotle and this entails not only that nothing can be really added to it but also that it is unrevisable. Kant's overall epistemological framework makes it particularly clear why (LE) seems so natural or even obvious. As Kant studies the conditions of possibility of knowledge, he sees logic as one of these conditions. With logic having this status, logical theses do not seem to be like the more mundane kinds of knowledge which we can change, as a change of logic would be fatal for the whole system of knowledge, would cause the collapse of the whole edifice. This is the gist of the exceptionalist view about logic.

It is possible to be an exceptionalist in the most strict way claiming that no new discoveries in logic are possible, yet this would be hardly plausible, perhaps even Kant could not embrace it. As improbable as Kant could consider it, there would still remain the possibility that we discover that a given law of inference which we considered as valid is not valid after all, as there are examples of true premises and a false conclusion of the requisite form. Or we can part company with Kant's thesis about completeness of logic and claim that we can subsequently discover that a law of inference which we did not consider at all or did not consider as logically valid, indeed is logically valid. With this view, one can still remain an exceptionalist in a sense. It is still a long path to a strongly anti-exceptionalist view that we can have more logics and choose between them according to some relevant criteria, i.e., revise our logic by swapping it for another one. Let us now examine anti-exceptionalism, as by means of the contrast we can understand also exceptionalism better.

2 Anti-exceptionalism

The opposite of exceptionalism is the thesis that logic should be regarded just as one of many scientific disciplines. Logical theories would then be assessed and potentially revised in the same way as all the others, for example when we adopt relativistic physics instead of Newtonian physics, when we adopt Darwinism instead of views not countenancing evolution of species, etc. In a similar manner, we can, for instance, adopt intuitionistic logic instead of classical logic. The term *anti-exceptionalism* has been recently

---

timmung des Vorgetragenen als Verbesserungen anrechnen will, welches aber mehr zur Eleganz, als zur Sicherheit der Wissenschaft gehört. Merkwürdig ist noch an ihr, daß sie auch bis jetzt keinen Schritt vorwärts hat tun können, und also allem Ansehen nach geschlossen und vollendet zu sein scheint. (Kant, 1954, B VII)

coined by Ole Hjortland. The view itself, though, has a longer tradition already, even though it is relatively young, given the venerable history of logic. Hjortland mentions two important contemporary anti-exceptionalists, namely Timothy Williamson and Graham Priest. The original source, however, is Quine of *Two dogmas*, as is appreciated by Williamson when he formulates the doctrine:

> We can use normal scientific standards of theory comparison in comparing the theories generated by rival consequence relations. Thus the evaluation of logics is continuous with the evaluation of scientific theories, just as Quine suggested. (Williamson, 2017, p. 14)

The various logical systems are thus seen partly as rivals, partly more generally as alternatives, yet in general not differing from other kinds of theories. The logical systems are then assessed as theories by a variety of relevant criteria. These include correspondence to data, even though in case of logic it is not straightforward to explain what exactly the data are and how they can be arrived at. Yet besides this obvious criterion, there is also, as we know already from Quine, simplicity, elegance, strength, explanatory power and the list could probably go on. This variety of criteria suggests that contexts are possible in which more theoretical choices are rational. Yet still, there are still the mentioned criteria and it by no means holds that anything goes.

Priest and Williamson discuss the respective merits of classical and paraconsistent logic and disagree about which one we should accept. They see the liar paradox as the neuralgic point which forces us to make such a decision. Both authors agree that in face of this paradox, either classical logic or the unrestricted truth predicate has to be sacrificed, yet they choose differently. While Priest thinks that restricting the truth predicate is an unsystematic and ad hoc step, and that paraconsistent logic scores better at capturing how we actually reason, Williamson thinks that classical logic still has such merits that it is irrational to abandon it. The key feature of anti-exceptionalism is that it emphasizes that in logic we have to make *decisions*, at least at some junctures of the development of the discipline. Ole Hjortland expresses some surprise that Williamson and Priest are both anti-exceptionalists and yet infer quite different conclusions from it, as Williamson is an ardent advocate of classical logic, while Priest argues in favour of paraconsistent logic. But if their shared anti-exceptionalism entails the necessity to make decisions regarding logic, then we should only expect that people will differ as to what decision they take. Now, it will be instructive to learn about another view which anti-exceptionalism is opposed to.

2.1 Against foundationalism

Anti-exceptionalism, despite the plurality of its specific forms, must be opposed to a view dubbed as *foundationalism* by Stewart Shapiro (1991). Foundationalists consider logic as the fundament of our whole knowledge and as such it cannot be changed without causing the collapse of the whole edifice resulting in irrationality. Refusing this kind of attitude clearly does not have to mean that we can change logic just according to our will or even whim. Such a change can still be difficult, yet not impossible, perhaps we have to take very specific measures in order to keep the edifice stable while engaging in the change.

The fact that logic is revisable might naturally surprise many people. If true, it would be a fact comparable to the discovery that we can change fundamental laws of physics or even geometry, as when non-Euclidian geometries were legitimized and Einsteinian physics superseded Newtonian physics. In case of physics, though, it is still not clear that we can really in any sense choose which theory we adopt. Maybe there is always just one correct path to go and when revising our physics, we have to adopt just the one new theory which is confirmed by the evidence we have at our disposal. In a similar vein, one could think that there is some kind of reality logic should model and this reality has to be matched as much as possible by our theories. In this sense any plurality of acceptable theories can at best be caused by lacking evidence to decide between the rival theories.

This view is surely a possible one, yet I do not consider it very plausible in case of any science, let alone in the case of logic. Thomas Kuhn has brought a lot of convincing examples of junctures in the history of natural sciences when the great minds did not just discover the way the reality is but also had to make important decisions which had lots of alternatives (Kuhn, 1962). And still, the idea of an independent reality which we study and which simply is there for us to discover is much more problematic in case of logic than in case of natural sciences. If the picture was shown to be implausible in case of natural sciences, then *a fortiori* we cannot stick with it in case of logic.

Note though that by opening the possibility to make significant choices regarding our theories, we therewith by no means claim that anything goes and that all choices are equal. Not only is it possible that some choices are significantly better than the others, the fact that there are choices to make explains what the greatness of great scientists consists in. The choices they made were often not anyhow evident, yet they have proved to be particularly felicitous. Still, this does not entail that there were not other choices, which would have been similarly good.

The debate we have about which theory to choose, be it in physics or in logic, can at least in some cases lead to weeding off of some clearly unacceptable options and still let more than one good choice open. Clearly, the debate can always be reopened for good reasons and some options which have so far seemed acceptable are discarded, while new options which were not on the table before can open up. I think that countenancing these possibilities makes for a particularly interesting version of anti-exceptionalism. We shall now proceed to examine how Quine, at different stages of his development, sometimes defended and sometimes attacked it.

3 Quine's conflict

It is overall not clear whether Quine contradicted himself regarding the dilemma about exceptionality of logic. The positions he expressed at different stages can to some degree be brought into harmony, although it would be a non-trivial undertaking. It is somewhat baffling that Quine did not address what seems to be a clear conflict in his views on logic.

3.1 Quine as an anti-exceptionalist

The locus classicus of Quine's anti-exceptionalism is obviously his famous article *Two Dogmas of Empiricism* in which he develops his holist epistemology which has become very influential. Anticipating what has to be documented by exact quotes, we can say that it is unclear what should be so new about the modern-day anti-exceptionalism, given how many of its salient features were already put forward by Quine.

The overall picture of how knowledge works put forward in *Two Dogmas* is that it is rather the wholes of our theories which are adjudicated and deemed right or wrong, not the individual statements. It remains somewhat open how we should delimit the wholes. In some sense all of our knowledge forms a whole, yet obviously it also has to be divided into relatively independent areas, such as that of mathematics. All these area are, nevertheless, always interconnected or rather they could be interconnected if we choose

Is Logic Exceptional?

to do so.

Quine still considers himself as an empirist which can be documented by how much he speaks about sensory experience, as in the following quote:

> Each man is given a scientific heritage plus a continuing barrage of sensory stimulation; and the considerations which guide him in warping his scientific heritage to fit his continuing sensory promptings are, where rational, pragmatic. (Quine, 1951, p. 43)

This quote closes the very article and indicates many of the features we would like to stress here and which seem to make Quine an anti-exceptionalist. Even though Quine is critical of the form empiricism had when he wrote the article, he can be seen rather as purporting to rectify than to abandon it. We can see that the *sensory promptings* still play a key role, although less straightforwardly than in the case of typical empiricism. Besides this, we see that he speaks about our *warping of the scientific heritage.* This is an acknowledgement both of the fact that we work in the context of a given tradition, yet at the same time we develop and modify it. While doing this, we can have some important choices to make.

Still, in many cases more than one decision regarding our theories can be good. This is closely tied to the fact that, where rational, these decisions are *pragmatic*. What this means is somewhat unclear and one would like to get more explanation from Quine. Yet we can guess that Quine wanted to oppose his pragmatism to an epistemology which would claim that knowledge is just about correspondence. According to such a view, there is simply some way the things out there are and our theories have to faithfully represent them. This naturally leads one to conclude that at any juncture, only one theoretical choice is right, namely that one which corresponds to how things happen to be.

Quine's holism, which sees the whole of our knowledge as one interconnected system of assertions entails that no assertion is in any strong sense exceptional, definitely not in the sense of unrevisability. Therefore, any proposition can, if we successfully adjust some other parts of the web of beliefs, be held true:

> Any statement can be held true come what may, if we make drastic enough adjustments elsewhere in the system. (Quine, 1951, p. 40)

On the other hand, any statement can be revised:

> Conversely, by the same token, no statement is immune to revision. Revision even of the logical law of the excluded middle has been proposed as a means of simplifying quantum mechanics. (Quine, 1951, p. 40)

Here we are left with no doubt that this anti-exceptionalism applies even to logic. It is, by the way, quite telling that Quine mentions explicitly logic, clearly with the presupposition that in case of logic the anti-exceptionalism will appear most problematic to some. Yet Quine does not hesitate here to embrace this position despite its controversial nature. Logic, according to him, is indeed not really exceptional as to how we evaluate and modify it, it is just one of our theories, logical statements belong to the web of our beliefs and can be adjusted just as any other part of the web. On the other hand, the web has its center or at least some relatively central parts and its periphery or maybe rather peripheries. Quine calls the periphery *sensory* at one place and we are therefore to understand that, unlike the central parts of the web, the peripheral ones are those which we are much more naturally prone to sacrifice in face of recalcitrant experience. The point, though, is that difference between the center and the periphery is merely one of degree, not one of kind. Regarding the relation of a given statement to possible experience, there are no definite condition of verification or confutation associated with a given statement:

> ...it is misleading to speak of the empirical content of an individual statement - especially if it be a statement at all remote from the experiential periphery of the field. (Quine, 1951, p. 40)

But the role of logical statements is not that straightforward, after all. Besides being in the web they also regulate the overall structure of the web, as is hinted at in the following quote:

> Re-evaluation of some statements entails re-evaluation of others, because of their logical interconnections - the logical laws being in turn simply certain further statements of the system, certain further elements of the field. Having re-evaluated one statement, we must re-evaluate some others, whether they be statements logically connected with the first whether they be statements of logical connections themselves. (Quine, 1951, p. 39)

Is Logic Exceptional?

The word *simply* does not seem to be appropriate. If they regulate the overall structure of the beliefs, logical statements cannot just simply be further statements, though this does not mean that we cannot construe a story about how the logical statement can both be in the web and beyond it. Such a story, though, will hardly be simple. Some authors have protested about the double role Quine ascribes to logic, particularly authors prone to redeem exceptionalism, for example Leech (2015) and Berger (2011). Obviously, Quine's anti-exceptionalism has its problems which we should keep in mind and deal with. Right now, though, before trying to examine how it can be completed, we should introduce ourselves to another line Quine develops which is hard to reconcile with his holism-based anti-exceptionalism.

3.2 Quine as an exceptionalist

Later during his career Quine came up with his famous *change of logic change of subject* thesis and a sketch of an argument for it. Again, as in *Two dogmas*, Quine presents quite a shocking thesis and, again just as in the previous case, offers only rather sketchy arguments for it. In fact, even much more sketchy than in the case of *Two dogmas*. Besides this, his new thesis seems to be quite opposite to his anti-exceptionalism and makes an impression of extreme exceptionalism. This is the passage which has become so famous:

> Here, evidently, is the deviant logician's predicament. When he tries to deny the doctrine, he only changes the subject. (Quine, 1986, p. 80)

Obviously, in *Two dogmas* Quine seemed to be very open to the possibility of changing a logic, while here it seems that we cannot even begin to endeavour such a change, any such activity is based on an original misunderstanding. We thus seem to be doomed to stick with the one logic we have. Quine, furthermore, thinks that it is the classical logic. At firth sight, though, his view does not appear to exclude the possibility that there could be two people one of whom would reason classically, while the other, say, intuitionistically and they could just not communicate their differences to each other, one being closed in his classical, the other in his intuitionistic reasoning.

3.3 Toleranzprinzip

Now it will be instructive to see how close Quine's views are to those presented by Carnap. Quine is known to have had both been strongly influenced by Carnap, yet also for breaking with some of the views advocated by logical positivists, among which Carnap belongs. *Two dogmas* are known as the text which most concisely criticizes the distinction between analytical and synthetic statements, so central for logical positivists in general and for Carnap in particular. His criticism, on the other hand, can be seen also a natural development of too crude a distinction towards more nuance. Quine's holism does not simply discard the distinction between analytic and synthetic statements, it rather claims that what what seemed to be a difference of kind is just a difference of degree, though sometimes of a great degree, as when you compare logical claims and claims about the color of houses in your street. His later exceptionalism about logic is, nevertheless, an example of line of though which also goes back to Carnap, even though not much attention has been paid to this fact.

Carnap formulated his views on plurality of logics and the possibility of changing a logic in *Logische Syntax der Sprache*. One of the particularly famous quotes is the following one:

> In logic, there are no morals. Everyone is at liberty to build his own logic, i. e. his own form of language, as he wishes. All that is required of him is that, if he wishes to discuss it, he must state his methods clearly, and give syntactical rules instead of philosophical arguments. (Carnap, 1934, p. 17)

Even though the formulation might be less explicit that in Quine, Carnap still seems to embrace the idea that changing a logic means changing a subject. When Carnap talks about building one's logic as about building a new form of language, he suggests that differences between logics are due to differences between languages. Otherwise put, trying to contradict a rival logic, means using a different language and therewith talking past the interlocutor. This interpretation is confirmed also by the proponents of logical pluralism who see Carnap as special kind of pluralist, namely as a *linguistic pluralist*, who claims that adopting a different logic means adopting a different language. This approach is contrasted with the so-called *non-linguistic pluralism*, for example of Beall and Restall.[3]

---

[3] This distinction between linguistic and non-linguistic pluralism was introduced by Restall

Is Logic Exceptional?

Given how much they share, it quite surprising that Carnap infers from this common core idea that differences in logic are due to differences of language quite a different conclusion than Quine. While Quine sees his thesis about the connection of logic and language as a strong support for logical monism and for futility or even non-sensicality of the idea of changing one's logic, Carnap sees it as a reason for a very radical form of logical pluralism. This is to be read already from the previous quote, but the following one is even more eloquent:

> The first attempts to cast the ship of logic off from the terra firma of the classical forms were certainly bold ones, considered from the historical point of view. But they were hampered by the striving after 'correctness'. Now, however, that impediment has been overcome, and before us lies the boundless ocean of unlimited possibilities. (Carnap, 1934, p. xv)

We can thus witness not only approval but even exulting at the idea of changing a logic. It is precisely because there can be no real disagreement between logics that we are free to simply choose our favourite logic. This position, by the way, again resembles some of the further tenets of Carnap's philosophy, namely his quest for demarcating what is a meaningful question, particularly in philosophy. As he believed that lots of the traditional philosophical questions were actually ill-conceived, so he thought the same about attempts at justifying a logic.

For example, Carnap (1928) argues that the dispute between idealism and realism does not make real sense, as neither of the two purported doctrines actually claims anything intelligible. In the same vein, according to Carnap, claiming that, e.g., intuitionistic logic is correct while classical logic is flawed is a similarly non-sensical thesis. In this sense Carnap opens up the space for alternative logics, though he certainly does not share the motivations of some of their key proponents, for example Brouwer who clearly maintained that intuitionistic logic is in a strong sense correct. This is exactly the *striving for correctness* Carnap denounces.

Overall, though, we should be puzzled by how opposite conclusions Carnap and Quine draw from their thesis that deviant logicians actually use different languages. I think that this divergence can be seen as argument that their shared thesis leads to an antinomy and thus should be refused.

---

himself in (Restall, 2002) and then reiterated in (Beall & Restall, 2006). Eklund (2017) provides more thorough classification of various kinds of logical pluralism, based on the mentioned linguistic/non-linguistic distinction.

4 Problems of the view

The common thesis of Carnap and Quine, which we can call CLCM (after Jarred Warren, who used this abbreviation meaning *Change of logic change of meaning*, see Warren, 2018), is thus rendered very problematic by showing that it leads to an antinomy between Quine's strict monism and Carnap's practically unrestricted pluralism. Yet in order to be truly untenable we should now show that both horns of the dilemma it forces on us are unacceptable.

The main problem with Carnap's view was sufficiently pointed out very early by Quine (1936). If the logical truths are merely truths by convention, Quine claimed, then it is unintelligible how we can do as much as to opt for one or another of the systems of conventions, say classical or intuitionistic logic, as we would be completely logic-free and thus unable to reason about the conventions we choose among. Carnap's view is thus too radical.

Furthermore, Carnap does not indicate which systems of inference rules should still be considered as a logic at all. Clearly, we have examples such as the classical logic or the intuitionistic logic. But certainly if we had a system in which, for example a disjunction of two formulae would always be equivalent to their conjunction, we would rather consider this as a nonsense than as an alternative logic. Besides this, it is unclear when two systems really differ, as Quine himself mentions that if somebody ascribed $\wedge$ all the properties of $\vee$ in classical logic and vice versa, we would consider this rather as a change of notation than a change of logic.

The problems of the Quinean monism might be little bit less clear and undisputable but still his view is hardly acceptable. As we already mentioned, Quine suggests that we simply have one or another logic and we cannot change it, yet it is entirely unclear how we could determine which logic it is. Quine believes it is classical logic but does not really argue for this.

Furthermore, Quine owes us an explanation of how the logical expressions acquire their meaning. Overall, Quine is sceptical towards the very notion of meaning of a given expression. He explicitly bans meanings from his ontological inventory in (Quine, 1948) and seems to allow only for the meaning of a theory as a whole. On the other hand, when he puts forward his CLCM thesis, he obviously relies on a very strong notion of meaning of a logical expression. We know at least that when we consider whether, say, negation and disjunction could work as in intuitionistic logic, i.e., not validating the law of the excluded middle, we already change the topic, i.e., we mean something else by these expressions than the classical logician means. Quine's pictures let's too much unexplained and it is unclear how it can be amended.

Overall, it becomes obvious that CLCM is too radical a thesis and we need to look for an alternative to it.

5 Looking for an alternative to CLCM

There are many ways you can look at the issue of changing a logic which we could consider as possible alternatives to CLCM. We begin by considering too radical a way of avoiding CLCM, namely to claim what is practically the exact opposite of it.

5.1 Alternative logicians always speak the same language?

As Quine claims that while talking about the validity of given logical law two logicians endow the logical expressions with different meanings, how about trying to claim the exact opposite? Maybe the two logicians in fact talk about the same thing and merely ascribe it different properties. Ultimately, the proposal would naturally continue, at most one of the logicians can be right, depending exactly on how it is with, say, disjunction and negation.

Such a position might seem *prima facie* plausible. Just as there is an independent physical reality about which physicists can have different opinions and dispute about them, so there maybe is an independent logical reality and various logicians propose their models of what this reality is like. I guess such a position can be maintained but with too many difficulties.

To begin with, one would have to, sooner or later, come up with an explanations of what this independent logical reality consists in. Is it some platonic realm of ideal logical entities? Or does it consist in how we think? Or in how we argue in everyday conversation? Furthermore, one would also have to explain how it is possible that so many theories about how this reality is keep being developed. Indeed, the number of purported logics is formidable and that would seem to entail that we are particularly unsure about this logical reality.

I do not find the platonistic alternative very intelligible and I think psychologism has been sufficiently refuted already by Frege (1884) and Husserl (1913). Yet even in the case of the last alternative, I remain puzzled as to what would be so interesting about the slightly differing intuitions language users have regarding the use of logical vocabulary that it would be worthwhile to come up with so many refined theories. In fact, people have conflicting intuitions about all kinds of vocabulary, not only about the logical vocabulary. One can always come up with a context in which we might be unsure about how to correctly use the words such as *dog, red, justice* and, yes, also such words as *not, or* and other logical expressions. It is, nevertheless, unclear what the benefits of adjudicating between possible divergent usages could be. Very often, the fairest thing to say is that it simply belongs to the rules guiding the use of the given expression that it remains ambiguous in certain contexts.

But besides this, we have to admit that Quine's CLCM thesis, although not acceptable in its entirety, does capture something important. We cannot say that when two logicians disagree about the validity of a single inference rule, they must be talking past each other, yet if they did not agree on any rules, guiding a logical expression such as *for all* or its formal counterpart $\forall$, then we would clearly have all the reason in the world to say that they talk about something else, and therefore past each other. When we strip the given expression of all inference rules it obeys in a given system of logic, for example in classical logic, there remains nothing which would justify calling it a meaningful expression, let alone a logical expression.

6 Steering clear of the two extremes

We have contrasted two opposite theses, the first one being CLCM, claiming that when two logicians disagree about a given logical law, they must be talking past each other, and the other one, to which CLCM can be seen as a reaction, claiming that the disputing logicians always talk about one and the same logical reality which is independent of us and of which inference rules we take to be valid. I want to redeem what seems to me valuable both in CLCM and in the opposite thesis. In principle, my proposal is a combination of both, with the extreme implications put aside.

First of all, even if a change of logic does not mean a straightforward change of subject, we can still partly bite the bullet of CLCM and maintain that when an intuitionist argues against the law, he partly changes the meaning of disjunction and negation or proposes to make such a change. Indeed, Quine himself, few pages after presenting the radical CLCM thesis, puts it with a much less radical tone:

Is Logic Exceptional?

> The intuitionist should not be viewed as controverting us as to the true laws of certain fixed logical operations, namely, negation and alteration[4]. He should be viewed rather as opposing our negation and alteration as unscientific ideas, and propounding certain other ideas, somewhat analogous, of his own. (Quine, 1986, p. 87)

Even though, shifting the meaning of logical expressions a little bit, the intuitionist obviously is not portrayed as doing something irrational in this passage. Yet still, Quine attempts to justify CLCM in as strong a form as possible and he invokes a variant of his charity principle. By interpreting somebody as contradicting our logic, we ascribe that person what we consider a wrong belief about logic, which Quine considers as extremely uncharitable. Jarred Warren, correctly, I think, warns us that this principle can lead us astray when used excessively. What he calls *charity trap* lures behind, namely the danger that the principle will render all disagreement impossible. The principle of charity can attain various forms and degrees. In a relatively moderate form it just says that one should not impute patently irrational beliefs on one's interlocutors, unless we discover serious and rare reasons for doing so, for example when we have to conclude that the interlocutor is mad. In such a form the principle can hardly be doubted.

On the other hand, one can take the principle too naively. The irrational beliefs are typically, of course, false. And one could therefore conclude that the charity principle imposes us never to ascribe false beliefs to our interlocutors. Or, more exactly put, what we consider as false beliefs. This would render any disagreement and therewith and real discussion impossible, which is why Warren calls such a conclusion *charity trap*. In full generality, the principle is therefore obviously absurd, yet it remains tempting to push it at least locally and claim that it holds in some specific area, for example in logic.

Thomas Kuhn (1962) in his celebrated *Structure of Scientific Revolutions* comes very close to such a view when it comes to physics or natural sciences in general. As interesting and insightful as his criticism of seeing the history of science as a mere accumulation of new knowledge is, the extreme charity claiming that we cannot consider those who claimed the Earth was flat as being wrong is untenable.

Quine tried to maintain this form of extreme charity in case of logic and

---

[4] By *alteration* Quine means disjunction here. In the quoted passage, he discusses the law of the excluded middle.

I do not see why it should become acceptable in this case. Disagreement, which by definition entails considering others as holding wrong beliefs, has to be possible in any meaningful discourse.

Yet still, we can maintain that by advocating a deviant logic, one does move the meanings of logical expression. The meaning, though, has to be changed only partially, in order not to genuinely change the subject, as Quine fears. Again, we can also agree with Kuhn that a change of scientific paradigm in general involves a change of meanings, yet this does not preclude disagreement and, at least in some cases, one side of the disagreement being right, the other being wrong, or at least less right. This is nicely put by the following quote from Hartry Field (what the quote states, though, is hardly surprising to anyone who read and accepted the argumentation of *Two dogmas*):

> On some readings of "differ in meaning", any big difference in theory generates a difference in meaning. On such readings, the connectives do indeed differ in meaning between advocates of the different all-purpose logics, just as 'electron' differs in meaning between Thomson's theory and Rutherford's; but Rutherford's theory disagrees with Thomson's despite this difference in meaning, and it is unclear why we shouldn't say the same thing about alternative all-purpose logics. (Field, 2009, p. 345)

Therefore, the truth is somewhere in between CLCM and the opposite thesis. Logicians can meaningfully disagree about logical laws, yet this disagreement has to be only partial, they can always dispute only about a very limited set of logical laws and only about some applications of these laws. This, then, is nothing special about logic, indeed it is not exceptional.

6.1 Anti-exceptionalism?

Most of my criticism has targeted CLCM which was used as an argument both for revisability and for unrevisability of logic. In a sense, therefore, for both anti-exceptionalism and for exceptionalism. But Quine's views from *Two dogmas* were not attacked very much. Does it mean that we embrace a version of anti-exceptionalism he proposes there and which, as we have argued, probably does not differ very much from what contemporary anti-exceptionalists such as Hjortland believe? Not quite exactly, though we will only briefly explain why.

Is Logic Exceptional?

What both exceptionalism and anti-exceptionalism have to share is the belief that we objectively have one or another specific logic among our web of beliefs, so that they can dispute whether it is a special part of the web or not. Yet I do not see why we should maintain that we have to ascribe to an individual person or to a given community one logic rather than another. The differences between serious logical systems such as classical or intuitionistic logic typically do not come to the fore in most of the everyday reasoning. The talk of having and changing one's logic engenders the impression of opening a new chapter in one's life, as if one decided to start eating less sugar or being more nice to others. Yet it is unclear how to recognize that a given person genuinely reasons classically and therefore also that she has converted to intuitionism. If having a logic is a misguided notion, then so is changing it.

In this sense anti-exceptionalism is a flawed view of how logical systems interact with the daily practice of reasoning. Exceptionalism, as should be clear from my arguments, does not fare any better. We need a fundamentally different approach which yet has to be developed.

References

Beall, J., & Restall, G. (2006). *Logical Pluralism.* Oxford: Oxford University Press.
Berger, A. (2011). Kripke on the incoherency of adopting a logic. In A. Berger (Ed.), *Saul Kripke* (pp. 177–210). Cambridge University Press.
Carnap, R. (1928). *Der logische Aufbau der Welt.* Berlin-Schlachtensee, transl. The Logical Structure of the World, Berkeley: University of California Press, 1967.
Carnap, R. (1934). *Logische Syntax der Sprache.* Wien: Springer, transl. The Logical Syntax of Language, London: Kegan, Paul, Trench Teubner & Cie, 1937.
Eklund, M. (2017). Making sense of logical pluralism. *Inquiry*, 1–22.
Field, H. (2009). Pluralism in logic. *Review of Symbolic Logic*, 2(2), 342–359.
Frege, G. (1884). *Die Grundlagen der Arithmetik.* Breslau: Wilhelm Koebner.
Hjortland, O. (2017). Anti-exceptionalism about logic. *Philosophical Studies*, *174*(3), 631–658.

Husserl, E. (1913). *Logische Untersuchungen*. Halle: Niemeyer.
Kant, I. (1954). *Kritik der reinen Vernunft*. Hamburg: Felix Meiner.
Kuhn, T. (1962). *The Structure of Scientific Revolutions*. University of Chicago Press.
Leech, J. (2015). Logic and the laws of thought. *Philosophers' Imprint*, *15*(12), 1–27.
Quine, W. V. O. (1936). Truth by convention. *Journal of Symbolic Logic*, *1*(1), 77–106.
Quine, W. V. O. (1948). On what there is. *Review of Metaphysics*, *2*(1), 21–38.
Quine, W. V. O. (1951). Two dogmas of empiricism. *Philosophical Review*, *60*(1), 20–43.
Quine, W. V. O. (1986). *Philosophy of Logic*. Cambridge: Harvard University Press.
Restall, G. (2002). Carnap's tolerance, language change and logical pluralism. *Journal of Philosophy*, *99*(8), 426–443.
Shapiro, S. (1991). *Foundations without Foundationalism*. Oxford: Oxford University Press.
Sher, G. (1991). *The Bounds of Logic: a Generalized Viewpoint*. Cambridge, MA: MIT Press.
Warren, J. (2018). Change of logic, change of meaning. *Philosophy and Phenomenological Research*, *96*(2), 421–442.
Williamson, T. (2017). Semantic paradoxes and abductive methodology. In B. Armour-Garb (Ed.), *Reflections on the Liar*. Oxford University Press.

Pavel Arazim
Czech Academy of Sciences, Institute of Philosophy
The Czech Republic
E-mail: arazim@flu.cas.cz

The Theory of Topic-Sensitive Intentional Modals

FRANCESCO BERTO[1]

Abstract: A Topic-Sensitive Intentional Modal (TSIM) is a two-place, variably strict modal with an aboutness or topicality constraint, of the form '$X^\phi \psi$' (read: 'Given ϕ, the agent X's that ψ', X being some mental state or act). TSIMs do nice things for mainstream and formal epistemology, belief revision theory, and mental simulation theory. I present a basic formal semantics for TSIMs and explore three readings of '$X^\phi \psi$' one gets by imposing different constraints on their truth conditions: (1) as expressing knowability relative to information ('Given total information ϕ, one is in the position to know that ψ'), inspired by Dretske's view that what one can know depends on the available (empirical) information; (2) as a mental simulation operator ('In mental simulation starting with input ϕ, one imagines that ψ') capturing features of mainstream mental simulation theories, like that of Nichols and Stich; (3) as a hyperintensional belief revision operator ('After (statically) revising by ϕ, one believes that ψ'), reducing the idealization of cognitive agents one finds in standard doxastic logics and AGM. I close by mentioning developments of TSIM theory currently in progress.

Keywords: intentionality, epistemic and doxastic logic, aboutness theory, hyperintensionality, mental simulation, belief revision, knowabilty, information

1 Introduction

We have learned since (Hintikka, 1962) how to treat notions like *knows, believes, is informed that* using modal logic: we interpret them as quanti-

---

[1] This paper is published within the project 'The Logic of Conceivability', funded by the European Research Council (ERC CoG), Grant Number 681404. It summarizes material published or forthcoming in (Berto, 2017a, 2018; Berto & Hawke, 2018), all of which are Open Access publications falling under the terms of the Creative Commons Attribution 4.0 International License (http://creativecommons.org/licenses/by/4.0/). Several people are thanked in those papers. The Logic of Conceivability gang deserves special rehearsed thanks – in particular Chris Badura, Ilaria Canavotto, Jorge Ferreira, Peter Hawke, Karolina Krzyzanowska, Aybüke Özgün, Tom Schoonen, Anthi Solaki. Thanks a lot to the organizers of the *Logica* conference for inviting me to present this material, and to the audience at *Logica 2018* for helpful comments and remarks.

fiers over possible worlds, restricted from the viewpoint of a given world by an accessibility relation (hopefully) endowed with some intuitive meaning. '$X\phi$' ('The agent Xs that ϕ') is true at w just in case ϕ is true at a bunch of worlds accessible via the relation R from w. By imposing simple conditions on R, we then validate various principles characteristic of different modal systems. Some conditions are more contentious than others. We all agree that R should be reflexive for 'X' to be read as 'knows', it shouldn't for it to be read as 'believes'. But we debate on whether R should be transitive, for we disagree on whether Positive Introspection should hold for knowledge or belief: does Xing that ϕ entail that one Xs that one Xs that ϕ?

All of this is well known. The rehearsal just provided is in order to highlight the three main ways in which the very general framework for intentional operators I want to sketch in *this* paper differs from the mainstream tradition. Call such framework the theory of Topic-Sensitive Intentional Modals (TSIM – read it as '*ZIMM!*'):

(1) The Hintikkan operators are one-place modals. The TSIMs are two-place modals: things of the form '$X^\phi\psi$', to be generically read as 'Given ϕ, the agent Xs that ψ', where X is some mental state or act.

(2) The $X^\phi\psi$'s are *variably strict* modals. Variability represents the contextual selection of information the agent imports into the Xed content on the basis of ϕ. The operators turn out to be nonmonotonic: epistemic logic, in TSIM clothing, becomes a kind of conditional logic.

(3) The $X^\phi\psi$'s encompass a *topicality* or *aboutness* filter capturing their standing for *intentional* mental states: states which are directed towards, or are about, a certain content or topic represented in the mind.

Ideas (1) and (2) are in the literature: two-place epistemic or doxastic operators expressing conditional belief, or static and dynamic belief revision ('$B^\phi\psi$': 'Conditional on ϕ, one believes ψ'; '$[\phi]\psi$': 'After revising one's beliefs by ϕ, it is the case that ψ') have been explored, e.g., in Dynamic Epistemic Logic and in modal recaptures of AGM (Spohn, 1988; Segerberg, 1995; Lindström & Rabinowicz, 1999; Board, 2004; van Ditmarsch, 2005; Asheim & Sövik, 2005; Leitgeb & Segerberg, 2005; van Benthem, 2007; Baltag & Smets, 2008; van Ditmarsch, van der Hoek, & Kooi, 2008; van Benthem, 2011; Girard & Rott, 2014; etc.).

Idea (3) is relatively new, though variously related to work on tautological or analytic entailment (Parry, 1933; van Fraassen, 1969; Angell, 1977;

The Theory of Topic-Sensitive Intentional Modals

Fine, 1986; Correia, 2004; Ferguson, 2014) and awareness logic (Fagin & Halpern, 1988; Schipper, 2015). So let me comment on (3) a bit.

Aboutness is 'the relation that meaningful items bear to whatever it is that they are *on* or *of* or that they *address* or *concern*' (Yablo, 2014, p. 1). Research on aboutness is burgeoning: (Fine, 2014, 2016; Hawke, 2017). Such works have clarified that what a sentence is about can be (properly) included in what another one is about. Thus, to the extent that aboutness is a component of content, contents should be capable of standing in mereological relations (Yablo, 2014, Section 2.3), (Fine, 2016, Sections 3-5). They should be capable of being fused into wholes which inherit the proper features from the parts (Yablo, 2014, Section 3.2).

Yablo and Fine address aboutness mainly as a feature of linguistic representations, but Chapter 7 of (Yablo, 2014) gets into the aboutness of epistemic states. And rightly so, because another kind of representation bears aboutness, too: mental representation. Maybe Brentano was wrong when he said that all mental states bear intentionality, but some do, and 'every intentional state or episode has an object – something it is about or directed on' (Crane, 2013, p. 4).

The insight behind TSIM theory is that we should take at face value the view of belief, knowledge, (cognitive) information, but also of other notions less explored in formal logic, like imagination and mental simulation, as (propositional) representational mental states bearing intentionality, that is, being about states of affairs, issues, situations, or circumstances which make for their contents. I will generically call these things *topics*, and provide a simple formal mereology for them. The semantics for our TSIMs will be given in a kind of conditional logic framework, with an added mereology of topics.

Besides being nonmonotonic thanks to (2), our $X^\phi \psi$'s will turn out to be *hyperintensional*, differentiating between necessarily or logically equivalent contents, thanks to their topicality or aboutness filter (3). And rightly so, because *thought* is hyperintensional. Our mental states – believing, supposing, desiring, hoping, fearing – can treat logically or necessarily equivalent contents differently: Lois Lane can wish that Superman is in love with her without wishing that Clark Kent is in love with her, although (if Barcan Marcus and Kripke are right) it is metaphysically impossible for Superman to be other than Clark Kent. We can think that $75 \times 12 = 900$ without thinking that Fermat's Last Theorem is true. But given the necessity of mathematical truths, the two make for the same content or proposition in possible worlds semantics: the total set of worlds.

One further feature of TSIM theory brings back continuity with the standard Hintikkan framework: starting from a basic semantics for our $X^\phi\psi$'s, which I will present in Section 2 below, one can add constraints on the accessibility relations (or, as we will see, functions) used in their truth conditions. Such constraints validate different logical consequences or principles involving the operators, and come with different interpretations for them. (Just as, starting from **K** as our basic normal modal logic, we get stronger systems by adding constraints on accessibility, and the different principles characteristic of **B**, **S4**, etc., come with different interpretations of the relevant modals.)

In Section 3, I will give an overview of three such interpretations; not because they are especially good, or because they are the only ones available, but just because, as a matter of fact, these are the ones I have explored in various works, alone or with friends:

Section 3.1 : '$X^\phi\psi$', relabeled as '$K^\phi\psi$', as expressing a notion of *knowability relative to information* ('Given total information ϕ, one is in the position to know that ψ'), inspired by Dretske's (1999) view that what an agent can know is dependent on the available (empirical) information. Peter Hawke and I have developed this in a paper forthcoming in *Mind* (Berto & Hawke, 2018).

Section 3.2: '$X^\phi\psi$', relabeled as '$I^\phi\psi$', as expressing an *imagination* or *mental simulation* operator ('In an act of imagination starting with input ϕ, one imagines that ψ'), capturing ideas found, e.g., in mental simulation theories from cognitive science like (Nichols & Stich, 2003), and in Williamson's (2007) imagination-based modal epistemology. I have presented this in a paper that has come out in *Philosophical Studies* (Berto, 2017a).

Section 3.3: '$X^\phi\psi$', relabeled as '$B^\phi\psi$', as expressing a hyperintensional *conditional belief*, or (static) *belief revision* operator ('Conditional on ϕ, one believes ψ', or: 'After revising by ϕ, one believes ψ'), which reduces the logical idealization of cognitive agents affecting similar operators in standard doxastic logics as well as in AGM. I have presented this in a paper that has come out in *Erkenntnis* (Berto, 2018).

All of the above are mere initial explorations of the TSIM world. In the conclusive Section 4, I briefly speak of possible further work and of how others are currently developing some TSIM ideas.

The Theory of Topic-Sensitive Intentional Modals

2 The basic semantics

Take a propositional language $\mathcal{L}$ with an indefinitely large set $\mathcal{L}_{AT}$ of atomic formulas, p, q, r ($p_1, p_2, \ldots$), negation $\neg$, conjunction $\wedge$, disjunction $\vee$, a strict conditional $\prec$, X standing for a generic TSIM, round parentheses as auxiliary symbols (,). I use $\phi, \psi, \chi, \ldots$, as metavariables for formulas of $\mathcal{L}$. The well-formed formulas are items in $\mathcal{L}_{AT}$ and, if ϕ and ψ are formulas:

$$\neg \phi \mid (\phi \wedge \psi) \mid (\phi \vee \psi) \mid (\phi \prec \psi) \mid X^\phi \psi$$

Outermost brackets are usually omitted. We identify $\mathcal{L}$ with the set of its well-formed formulas. In the metalanguage I use variables $w, w_1, w_2, \ldots$, ranging over worlds, x, y, z ($x_1, x_2, \ldots$), ranging over *topics* (I'll say more on these in a minute), and the symbols $\Rightarrow, \Leftrightarrow, \&, or, \sim, \forall, \exists$, read the usual way. A *frame* for $\mathcal{L}$ is a tuple $\mathfrak{F} = \langle W, \{R_\phi \mid \phi \in \mathcal{L}\}, \mathcal{T}, \oplus, t \rangle$, understood as follows:

- W is a non-empty set of possible worlds.

- $\{R_\phi \mid \phi \in \mathcal{L}\}$ is a set of accessibilities between worlds, where each $\phi \in \mathcal{L}$ has its own $R_\phi \subseteq W \times W$. These may satisfy a number of different conditions, to which I come in a minute.

- $\mathcal{T}$ is a set of *topics*. We may understand topics as the abstract or concrete situations (the configurations of objects and properties), or issues, or Yablovian or Finean subject matters the formulas of $\mathcal{L}$ involved in intentional ascriptions are *about*. (We need no more for our propositional logic purposes. In particular, we can stay silent on how they may be interpreted: as certain divisions of the set of worlds, as Finean truthmakers, structured entities, or else. We only ask them to obey the mereological constraints coming next.)

- $\oplus$ is *topic fusion*, a binary operation on $\mathcal{T}$ making of topics part of larger topics and satisfying, for all $x, y, z \in \mathcal{T}$:

    – (Idempotence) $x \oplus x = x$
    – (Commutativity) $x \oplus y = y \oplus x$
    – (Associativity) $(x \oplus y) \oplus z = x \oplus (y \oplus z)$

    Fusion shall be unrestricted: $\oplus$ is always defined on $\mathcal{T}$: $\forall xy \in \mathcal{T}$ $\exists z \in \mathcal{T}(z = x \oplus y)$. *Topic parthood*, $\leq$, can then be defined the usual way: $\forall xy \in \mathcal{T}(x \leq y \Leftrightarrow x \oplus y = y)$. Thus, it's a partial ordering – for all $x, y, z \in \mathcal{T}$:

- (Reflexivity) $x \leq x$
- (Antisymmetry) $x \leq y$ & $y \leq x \Rightarrow x = y$
- (Transitivity) $x \leq y$ & $y \leq z \Rightarrow x \leq z$

Then $\langle \mathcal{T}, \oplus \rangle$ is a join semilattice (we could have done things the other way around, having the partial ordering in the frames and defining fusion out of it, but this would have made little difference, and the algebraic setting might be more intuitive). We may also assume that $\mathcal{T}$ is complete: any set of topics $S \subseteq \mathcal{T}$ has a fusion $\oplus S$. Finally, we can think of all topics in $\mathcal{T}$ as built via fusions out of *atoms*, topics with no proper parts ($Atom(x) \Leftrightarrow \sim \exists y(y < x)$, with $<$ the strict order defined from $\leq$) which we stipulate to be at the bottom of our semilattice. $\langle \mathcal{T}, \oplus \rangle$ is needed to assign topics to formulas of $\mathcal{L}$, as follows.

- $t : \mathcal{L}_{AT} \to \mathcal{T}$ is a function, such that if $p \in \mathcal{L}_{AT}$, then $t(p) \in \{x \in \mathcal{T} | Atom(x)\}$: atomic topics are assigned to atomic formulas (this makes of our $\mathcal{L}$ an idealized language: grammatically simple sentences of ordinary language can be about intuitively complex topics). t is extended to the whole of $\mathcal{L}$. If the set of atoms in ϕ is $\mathfrak{At}\phi = \{p_1, \ldots, p_n\}$, then:

  - $t(\phi) = \oplus \mathfrak{At}\phi = t(p_1) \oplus \ldots \oplus t(p_n)$.

A formula is about what its atoms, taken together, are about.

This mereology of topics (of which a more refined version is Peter Hawke's *issue-based theory*: see Hawke, 2017) will allow our TSIMs to make hyperintensional distinctions. However, we don't get as fine-grained as the syntax of $\mathcal{L}$. By induction on the construction of formulas, $t(\phi) = t(\neg\neg\phi)$ (recall Frege on the *Sinn*-preservation of Double Negation). Also, $t(\phi) = t(\neg\phi)$: a formula is about what its negation is about (no matter how we understand the topic of the whiteness of snow, 'Snow is white' is about that, and that is what 'Snow is not white' is also about). And not only $t(\phi \wedge \psi) = t(\phi \wedge \psi)$, but also, e.g., $t(\phi \wedge \psi) = t(\phi) \oplus t(\psi) = t(\phi \vee \psi)$. In the literature, these are often taken as key requirements for a good recursive

The Theory of Topic-Sensitive Intentional Modals

account of aboutness or subject matter (see Yablo, 2014, p. 42; Fine, 2016, p. 1).

A *model* $\mathfrak{M} = \langle W, \{R_\phi \mid \phi \in \mathcal{L}\}, \mathcal{T}, \oplus, t, \Vdash \rangle$ is a frame with an interpretation $\Vdash \subseteq W \times \mathcal{L}_{AT}$, relating worlds to atoms: we read '$w \Vdash p$' as meaning that p is true at w, '$w \nVdash p$' as $\sim w \Vdash p$. $\Vdash$ is extended to all formulas of $\mathcal{L}$ thus:

- (S¬) $w \Vdash \neg\phi \Leftrightarrow w \nVdash \phi$
- (S∧) $w \Vdash \phi \wedge \psi \Leftrightarrow w \Vdash \phi \ \& \ w \Vdash \psi$
- (S∨) $w \Vdash \phi \vee \psi \Leftrightarrow w \Vdash \phi \ or \ w \Vdash \psi$
- (S≺) $w \Vdash \phi \prec \psi \Leftrightarrow \forall w_1 (w_1 \Vdash \phi \Rightarrow w_1 \Vdash \psi)$
- (SX) $w \Vdash X^\phi \psi \Leftrightarrow$ **(1)** $\forall w_1 (w R_\phi w_1 \Rightarrow w_1 \Vdash \psi) \ \& \ $ **(2)** $t(\psi) \leq t(\phi)$

For '$X^\phi \psi$' to come out true at w we ask, thus, for two things to happen:

(1) ψ must be true at all worlds w_1 one looks at, via the accessibility determined by ϕ (more specific readings of '$wR_\phi w_1$' will come in Section 3: these depend on the conditions we add). This is the truth-conditional component making of $X^\phi \psi$ a variably strict quantifier over worlds.

(2) ψ must be fully on topic with respect to ϕ. This is the aboutness-preservation component.

(SX) can be equivalently expressed using set-selection functions (Lewis (1973), pp. 57-60). Each $\phi \in \mathcal{L}$ comes with a function $f_\phi : W \to \mathcal{P}(W)$ outputting the set of accessible worlds, $f_\phi(w) = \{w_1 \in W \mid wR_\phi w_1\}$. If $|\phi| = \{w \in W \mid w \Vdash \phi\}$, we can rephrase the clause for X as:

- (SX) $w \Vdash X^\phi \psi \Leftrightarrow$ **(1)** $f_\phi(w) \subseteq |\psi| \ \& \ $ **(2)** $t(\psi) \leq t(\phi)$

The two formulations are equivalent as $wR_\phi w_1 \Leftrightarrow w_1 \in f_\phi(w)$. However, either formulation is at times handier than the other. In particular, we will phrase the additional conditions on the semantics of our TSIMs in Section 3 using the f's.

Finally, logical consequence is truth preservation at all worlds of all models. With Σ a set of formulas:

$\Sigma \vDash \psi \Leftrightarrow$ in all models $\mathfrak{M} = \langle W, \{R_\phi \mid \phi \in \mathcal{L}\}, \mathcal{T}, \oplus, t, \Vdash \rangle$ and for all $w \in W$: $w \Vdash \phi$ for all $\phi \in \Sigma \Rightarrow w \Vdash \psi$

For single-premise entailment, I write $\phi \vDash \psi$ for $\{\phi\} \vDash \psi$. Logical validity, $\vDash \phi$, truth at all worlds of all models, is $\emptyset \vDash \phi$, entailment by the empty set of premises.

The logic induced by the semantics for the extensional operators is just classical propositional, with $\prec$ a strict S5-like conditional (i.e., one equivalent to the necessitation of a material conditional, where the relevant necessity is S5). The novelty comes with $X^\phi \psi$, whose logical behavior we are now going to unpack.

3 Adding conditions

One can impose different conditions on the f's:

(C0) $|\phi| \subseteq f_\phi(w)$

(C1) $f_\phi(w) \subseteq |\phi|$

(C2) $|\phi| \neq \emptyset \Rightarrow f_\phi(w) \neq \emptyset$

(C3) $f_\phi(w) \subseteq |\psi|$ & $f_\psi(w) \subseteq |\phi| \Rightarrow f_\phi(w) = f_\psi(w)$

(C4) $f_\phi(w) \cap |\psi| \neq \emptyset \Rightarrow f_{\phi \wedge \psi}(w) \subseteq f_\phi(w)$

The three interpretations of our TSIMs to be explored now come, respectively, from (1) adding (C0), (2) adding (C1) (and, tentatively, (C3)), and (3) imposing a total ordering on W (read as comparative plausibility in a belief system) that automatically validates (C1)-(C4). In each case, we restrict our attention to models that satisfy the relevant conditions. In each of the three subsections, I will only explore *some* notable validities and invalidities involving the TSIMs. There are many more, for which I refer to the source papers mentioned in Section 1 above.

3.1 Knowability relative to information

(C0) says that all the ϕ-worlds are selected, but allows for selected worlds which are not ϕ-worlds. With this one in place, we relabel our '$X^\phi \psi$' as '$K^\phi \psi$' and read it as expressing the *Knowability of ψ, Relative to Information ϕ* (KRI). This comes from Dretske (1999), who stressed the view that

The Theory of Topic-Sensitive Intentional Modals

knowledge depends on the (empirical) information available to us, where the role of incoming information is to narrow down the set of epistemically viable alternatives.[2] Thus, we read the accessibility $wR_\phi w_1$ as: 'Relative to w, w_1 is epistemically accessible on the basis of total information that ϕ', or: 'Relative to w, w_1 is not ruled out by knowledge based on the total information that ϕ'.

Information (1) eliminates possibilities, just as the truth of a meaningful sentence is, in general, compatible with some possibilities and not others; and (2) is about something, just as a meaningful sentence has a subject matter that it addresses. Knowability of ψ is, then, determined by the available information ϕ twice over: (1) once via the worlds ϕ makes epistemically accessible (that's the truth-conditional component of TSIMs), and (2) once via the topic ϕ is concerned with (that's the aboutness component).

A first simple validity comes via (C0) (the proof is trivial) and captures the idea that knowledge is factive:

(**Factivity**) $\{K^\phi \psi, \phi\} \vDash \psi$

When ψ is knowable based on the information that ϕ, and ϕ is true, ψ must be true as well. Notice that ϕ *needn't* be true: one point of departure of KRI from Dretske's view, is that Dretske takes all information to be veridical, whereas KRI is neutral on this, as per the (trivially proved) invalidity:

$K^\phi \psi \nvDash \phi$

(On the debate concerning the factivity of information, see, e.g., Floridi, 2015. Floridi himself is in favour. In the literature on belief revision, however, a weaker sense of information is often adopted, whereby (declarative) information is meaningful data, not perforce truthful. This is connected to what is sometimes called 'soft information', see, e.g., van Benthem, 2011; van Benthem & Smets, 2015.)

KRI, as well as all the other TSIMs, is closed with respect to conjunction elimination:

---

[2] The standard Hintikkan framework already embeds the impulse to parametrize knowledge to information: it models agent a's epistemic situation as a set of possible worlds, most straightforwardly understood as a's information or knowledge. Ascriptions $K_a \varphi$ are then naturally understood as capturing what is *knowable on this basis*. Various proposed readings draw out the conditionality. Consider the preferred interpretation in (Hintikka, 1962): $K_a p$ means roughly 'Relative to her knowledge, a is permitted to infer p'. Or consider a purely descriptive interpretation raised in Sect. 2.10 of (Hintikka, 1962): 'It follows from what a knows that p'.

(**Simplification**) $K^\phi(\psi \wedge \chi) \vDash K^\phi \psi \quad K^\phi(\psi \wedge \chi) \vDash K^\phi \chi$

Proof. We do the first one (for the second, replace ψ with χ appropriately). Let $w \Vdash K^\phi(\psi \wedge \chi)$. By (S$X$), for all w_1 such that $wR_\phi w_1$, $w_1 \Vdash \psi \wedge \chi$, thus by (S$\wedge$), $w_1 \Vdash \psi$. Also, $t(\psi \wedge \chi) = t(\psi) \oplus t(\chi) \leq t(\phi)$, thus $t(\psi) \leq t(\phi)$. Then, by (SX) again, $w \Vdash K^\phi \psi$. $\square$

The 'tracking' notion of knowledge due to Nozick (1981) does not necessitate that one who knows a conjunction is positioned to know the conjuncts. According to Kripke (2011a), this is a damning defect for Nozick's approach. KRI is free from such a defect. The companion of Simplification is:

(**Adjunction**) $\{K^\phi \psi, K^\phi \chi\} \vDash K^\phi(\psi \wedge \chi)$

Proof. Let $w \Vdash K^\phi \psi$ and $w \Vdash K^\phi \chi$, that is, by (SX): for all w_1 such that $wR_\phi w_1$, $w_1 \Vdash \psi$ and $w_1 \Vdash \chi$, so by (S$\wedge$) $w_1 \Vdash \phi \wedge \psi$. Also, $t(\psi) \leq t(\phi)$ and $t(\chi) \leq t(\phi)$, thus $t(\psi) \oplus t(\chi) = t(\psi \wedge \chi) \leq t(\phi)$. Then, by (S$X$) again, $w \Vdash K^\phi(\psi \wedge \chi)$. $\square$

All of the TSIMs explored in this paper share two further core features: (a) they are nonmonotonic, and (b) they display their hyperintensionality by invalidating, among other things, Closure under strict implication. Thus, in particular, for KRI:

(**Monotonicity**) $K^\phi \psi \nvDash K^{\phi \wedge \chi} \psi$

Countermodel. Let $W = \{w, w_1\}$, w R_p-accesses nothing, $wR_{p \wedge r} w_1$, $w_1 \nVdash q$, $t(p) = t(q) = t(r)$. Then $w \Vdash K^p q$, but $w \nVdash K^{p \wedge r} q$. $\square$

(**Closure under** $\prec$) $\{K^\phi \psi, \psi \prec \chi\} \nvDash K^\phi \chi$

Countermodel. Let $W = \{w, w_1\}$, $wR_p w_1$, $w \nVdash q$, $w_1 \Vdash q$, $w_1 \Vdash r$, $t(p) = t(q) \neq t(r)$. Then $f_p(w) \subseteq |q|$ and $t(q) \leq t(p)$, thus by (SX), $w \Vdash K^p q$. Also, $|q| \subseteq |r|$, thus by (S$\prec$), $w \Vdash q \prec r$. But although $f_p(w) \subseteq |r|$, $t(r) \nleq t(p)$, thus $w \nVdash K^p r$. $\square$

(a) Failure of Monotonicity comes from the TSIMs' variable strictness: $f_\phi(w)$ can differ from $f_{\phi \wedge \chi}(w)$. In particular, for KRI: the addition of new information may reduce one's knowledge. (b) Failure of Closure comes from the fact that $\prec$ can take one off-topic, whereas information is topic-sensitive: although all the ψ-worlds are χ-worlds, thus all the ϕ-selected

ψ-worlds are χ-worlds, ϕ may be information about the topic of ψ yet not be information about the topic of χ. Thus, given ϕ, one can come to know ψ but not χ even if there just is no way for ψ to be true while χ is not.

In the *Mind* paper with Peter, I argued that *both* the failure of Monotonicity *and* that of Closure help with the Kripke-Harman Dogmatism Paradox (Harman, 1973; Kripke, 2011b), whereby knowing agents seem to be immune to rational persuasion with new evidence. Suppose that agent x knows φ on the basis of information I_1. Suppose that evidence e, when received, disconfirms φ. Now, φ entails that $\neg(e \wedge \neg\varphi)$: hence φ entails that e, if true, is misleading evidence. So Closure yields that x knows that e (if true) is misleading on the question of φ. Now suppose that x receives new information I_2, positioning her to know e. In this case, by Monotonicity, x knows that e is misleading. In general, it seems that x can always ignore new countervailing evidence! Central strategies on the market deal with this paradox by targeting precisely Monotonicity (e.g., Harman, 1973) or Closure (e.g., Sharon & Spectre, 2010, 2017).

What of the venerable Platonic insight that knowledge as *epistéme* must, in some sense, be stable? It's captured by KRI's validating Transitivity. This is invalid for the other TSIMs explored below due to their variable strictness, but it holds for KRI thanks to (C0):

(**Transitivity**) $\{K^\phi\psi, K^\psi\chi\} \vDash K^\phi\chi$

Proof. Assume that $w \Vdash K^\phi\psi$ and $w \Vdash K^\psi\chi$. Thus: $\forall w_1(wR_\phi w_1 \Rightarrow w_1 \Vdash \psi)$ & $t(\psi) \leq t(\phi)$ and $\forall w_2(wR_\psi w_2 \Rightarrow w_2 \Vdash \chi)$ & $t(\chi) \leq t(\psi)$. Then $t(\chi) \leq t(\psi) \leq t(\phi)$. Further: by (C0), we have that $|\psi| \subseteq f_\psi(w)$ and, by (SX), that $f_\psi(w) \subseteq |\chi|$. Thus, $|\psi| \subseteq |\chi|$. Now, by (SX) again, we have that $f_\phi(w) \subseteq |\psi|$. Hence, $f_\phi(w) \subseteq |\chi|$. □

Knowledge is stable in that old knowledge cannot be lost as new one is accumulated. The intuitive case for Monotonicity is that it captures the core idea of the stability of knowledge. KRI suggests a different hypothesis: knowledge is stable in that it respects Transitivity. Suppose χ is known on the basis of information ψ. And suppose that one's information is refined insofar as new information χ is received upon which knowledge of ψ can be based. Transitivity says that χ is still knowable: no knowledge is lost in the update from ψ to ϕ.

Failure of Closure helps with Cartesian skepticism (Dretske, 1970). One's ordinary empirical information, delivered via sensory perception, positions one to know mundane facts, e.g., that one has hands. Now, having hands is incompatible with being a bodiless brain-in-vat whose phenomenal experience is systematically misleading. Yet it seems implausible that ordinary empirical information eliminates the possibility of radical sensory deception.

What of the intuition that knowability is closed under deduction? KRI validates *Closure Over Known Implication and Topic*:

(COOKIT) $\{K_A B, K_A(B \prec C)\} \vDash K_A C$

Proof. Let $w \Vdash K_A B$ and $w \Vdash K_A(B \prec C)$. By the former and (SX), for all w_1 such that $wR_A w_1$, $w_1 \Vdash B$, and $t(B) \leq t(A)$. By the latter and (SK) again, for all w_1 such that $wR_A w_1$, $w_1 \Vdash B \prec C$. Thus for all w_1 such that $wR_A w_1$, $w_1 \Vdash C$. Also, $t(B \prec C) = t(B) \oplus t(C) \leq t(A)$, thus $t(C) \leq t(A)$. Thus by (SX), $w \Vdash K_A C$. □

The Cartesian threat is reduced for COOKIT: it allows that, on pain of circularity, deductions from one's mundane empirical knowledge cannot yield knowledge that the senses are reliable. But it assures that mundane information positions one to know every *mundane* consequence of that information.

3.2 Imagination as mental simulation

I used (C1), and tentatively added (C3), in (Berto, 2017a), a paper on imagination as mental simulation. (C1) has it that all the ϕ-selected worlds will be ϕ-worlds – worlds making ϕ true. With this constraint in place, we relabel our '$X^\phi \psi$' as '$I^\phi \psi$' and read it as 'Given input ϕ, one imagines ψ', or, less tersely, 'In an act of imagination starting with input ϕ, one imagines ψ'. We now read the accessibility $wR_\phi w_1$ as: 'w_1 is one of the worlds where things are as imagined (at w), starting with input ϕ'.

'Imagination' is highly ambiguous: we use it to refer to all sorts of intentional activities, from free mental wandering to daydreaming and hallucinating. What we want to model here, though, is the kind of imaginative exercise we engage in when we want to anticipate what will happen if such-and-so turns out to be the case ('What will I do if I can't pay my mortgage tomorrow?'), or when, counterfactually, we want to ascertain responsibilities ('Would he have been hit by the car, had the driver respected the speed limit?'). We simulate alternatives to reality in our mind, to explore what would and would not happen if they were realized. It is widely agreed in cognitive psychology as well as philosophy (Byrne, 2005; Kind & Kung,

The Theory of Topic-Sensitive Intentional Modals

2016; Markman, Klein, & Surh, 2009) that imagination as mental simulation is of epistemic value, not only to improve future performance (that's on the know-how side), but also to make contingency plans by successfully anticipating future outcomes, or by learning from mistakes via the consideration of alternative courses of action (that's on the know-that side).

Acts of imagination as mental simulation have a deliberate starting point, given by the initial input: we set out to target an explicit content, *that* ϕ. It has been generally acknowledged (Langland-Hassan, 2016; Van Leeuwen, 2016; Wansing, 2015; Williamson, 2016) that imagination, so understood, is episodic and voluntary in ways belief is not: you can imagine that all of London has been painted blue (and try to guess how Londoners would react to that), but, having overwhelming evidence of the contrary, you cannot make yourself believe it.

In their well-known cognitive model of mental simulation, Nichols and Stich (2003) have 'an initial premiss or set of premisses, which are the basic assumptions about what is to be pretended' (p. 24). This may be made up by the conceiver ('Now let us imagine what would happen if ...'), or it may be given as an external instruction (think of going through a novel and take the sentences you read as your sequential input). But also, we integrate the explicit input ϕ with background information we import, contextually, depending on ϕ and what we know or believe: once the initial input is in, Nichols and Stich (2003) claim, 'children and adults elaborate the pretend scenarios in ways that are not inferential at all', filling in the explicit instruction with 'an increasingly detailed description of what the world would be like if the initiating representation were true' (pp. 26-28).

The additional details come from our information base (Van Leeuwen, 2016, p. 95): we imagine the last meeting between Heathcliff and Catherine in *Wuthering Heights*. We represent Heathcliff dressed as an an Eighteenth-Century country gentleman, not as a NASA astronaut. The text of the novel never says this explicitly, nor do we infer this from the text via sheer logic. Rather, we import such information into the represented situation, based of what we know: we know that the story is temporally located in the Eighteenth Century, and we assume, lacking information to the contrary from the text, that Heathcliff is dressed as we know country gentlemen were dressed at the time. The variability of strictness of our TSIMs now accounts for the contextual selection of the information we import in an act of imagination when we integrate its explicit input.

Also, in reality-oriented mental simulation we do not indiscriminately import unrelated contents into the conceived scenarios: '[We require] that the world be imagined as it is *in all relevant respects*' (Kind, 2016, p. 153). What is imported is constrained by what is *on-topic* with respect to the input. This is, again, the job of our topic-preservation filter. Topicality is a distinguishing feature of reality-oriented mental simulation, as opposed to free-floating mental wandering: you know that Nuku'alofa is the capital of Tonga, but this is immaterial to your imagining Catherine and Heathcliff's adventures as per Brontë's book, in so far as such adventures do not involve Tonga at all. The story is not *about* that. So you will not, in general, import such irrelevant content in your scenario.

In this setting, a reflexivity principle holds, thanks to (C1) (the proof is trivial), securing that the initial input is always imagined:

(**Success**) $\models I^\phi \phi$

(I flag here that this may make the framework unsuitable to model cases of so-called 'imaginative resistance': see Gendler, 2000). On the other hand, lacking (C0), Factivity fails – and rightly so, for mental simulation isn't factive: I imagine Crispin Wright working in Stirling, ϕ, but I develop the scenario in my imagination by importing my background (false) belief that Stirling is in England (may the Scots forgive me). I imagine that Crispin works in an English city, $I^\phi \psi$. ϕ is true, but it doesn't follow that ψ is true, Crispin works in an English city.

Two disjunction-involving features that hold for all of our TSIMs deserve some comment in the imagination reading. Yablo's 'paradigm of non-inclusion', that is, of (classically valid) entailment which is not aboutness-preserving, is the entailment from a formula to a disjunction between it and something else. This needs to fail, in particular, for the aboutness of imagination. When one imagines in an act whose explicit input is ϕ, that ψ, one does not thereby imagine a disjunction between the latter and an unrelated χ. Intuitively enough, the mental simulator need not be aware of that disconnected χ at all. Thus we need, and we get, a failure of:

(**Addition**) $I^\phi \psi \nvDash I^\phi(\psi \vee \chi)$

Countermodel. Let $W = \{w, w_1\}$, $wR_p w_1$, $w_1 \Vdash q$, $t(p) = t(q) \neq t(r)$. Then $t(q) \leq t(p)$, so by (SX), $w \Vdash I^p q$. But $t(q \vee r) = t(q) \oplus t(r) \nleq t(p)$, thus $w \nVdash I^p(q \vee r)$. □

(Notice that the inference fails for the right reason: although $\psi \models \psi \vee \chi$, disjunction brings in irrelevant, alien content.)

The Theory of Topic-Sensitive Intentional Modals

The other disjunction-involving issue has to do with the fact that imagination generally under-determines its contents (this is true, I think, for intentional states by default). We imagine things vaguely, without this entailing that we imagine vague things. An often-used example: you imagine the crowded streets of New York and you think about a complex scenario involving cabs running around, people in restaurants, skyscrapers, etc. You do not imagine all the details, but you want the details to be there, so to speak. Although you do not imagine the city building by building, New York is not a vague object in your scenario, one with an objectively indeterminate number of buildings. Either the number of buildings of New York is odd, or it is even. But you do not imagine it either way. So we need, and we get, a failure of:

(**Distribution**) $I^\phi(\psi \vee \chi) \not\models I^\phi \psi \vee I^\phi \chi$

Countermodel. Let $W = \{w, w_1, w_2\}$, $wR_p w_1$, $wR_p w_2$, $w_1 \Vdash q$ but $w_1 \not\Vdash r$, $w_2 \Vdash r$ but $w_2 \not\Vdash q$, $t(p) = t(q) = t(r)$. Then by (S∨), $w_1 \Vdash q \vee r$ and $w_2 \Vdash q \vee r$, so for all w_x such that $wR_p w_x$, $w_x \Vdash q \vee r$. Also, $t(q \vee r) = t(q) \oplus t(r) \leq t(p)$, thus by (SX), $w \Vdash I^p(q \vee r)$. However, $w \not\Vdash I^p q$ and $w \not\Vdash I^p r$ for both q and r fail at some R_p-accessible world. Thus by (S∨), $w \not\Vdash I^p q \vee I^p r$. □

In the *Phil Studies* paper, I tentatively added Condition (C3) above. Let us look at it again:

(C3) $f_\phi(w) \subseteq |\psi|$ & $f_\psi(w) \subseteq |\phi| \Rightarrow f_\phi(w) = f_\psi(w)$

(C3) was labeled there as a 'Principle of Equivalents in Imagination': in the context of the interpretation of TSIMs as imagination operators, one reads it as saying that when all the selected ϕ-worlds make ψ true and vice versa, ϕ and ψ are equivalent in that, when we imagine either, we look at the same set of worlds. (C3) validates a nice Substitutivity principle for equivalents in imagination:

(**Substitutivity**) $\{I^\phi \psi, I^\psi \phi, I^\phi \chi\} \models I^\psi \chi$

Proof. Suppose $w \Vdash I^\phi \psi$, $w \Vdash I^\psi \phi$, $w \Vdash I^\phi \chi$. By (SX), these entail, respectively, (a) $f_\phi(w) \subseteq |\psi|$ and $t(\psi) \leq t(\phi)$, (b) $f_\psi(w) \subseteq |\phi|$ and $t(\phi) \leq t(\psi)$, (c) $f_\phi(w) \subseteq |\chi|$ and $t(\chi) \leq t(\phi)$. From (a) and (b) we get $f_\phi(w) = f_\psi(w)$ (by (C3)) and $t(\phi) = t(\psi)$ (by antisymmetry of topic parthood). From these and (c) we get $f_\psi(w) \subseteq |\chi|$ and $t(\chi) \leq t(\psi)$. Thus by (SX) again, $w \Vdash I^\psi \chi$. □

Substitutivity says that 'equivalents in imagination' ϕ and ψ can be replaced *salva veritate* as indexes in $I^{...}$. This seems right, in spite of the many hyperintensional distinctions we may draw in our mind. For suppose that *bachelor* and *unmarried man* are for you equivalent in imagination: you are so firmly aware of their meaning the same, that you cannot imagine someone being one thing without imagining him being the other ($I^\phi \psi$ & $I^\psi \phi$ entails $t(\phi) = t(\psi)$: equivalents in imagination are always about the same thing for the conceiving subject). Thus, $I^\phi \psi$, when you imagine that John is unmarried, you imagine that he is a bachelor, and $I^\psi \phi$, when you imagine that John is a bachelor, you imagine that he is unmarried. Suppose $I^\phi \chi$: as you imagine that John is unmarried, you imagine that he has no marriage allowance. Then the same happens as you imagine that he is a bachelor, $I^\psi \chi$.

I said that the addition of (C3) was 'tentative'. That's because it also validates a kind of Special Transitivity principle which has good instances, but, in the context of imagination, may face counterexamples:

(**Special Transitivity**) $\{I^\phi \psi, I^{\phi \wedge \psi} \chi\} \vDash I^\phi \chi$

Proof. Suppose (a) $w \Vdash I^\phi \psi$ and (b) $w \Vdash I^{\phi \wedge \psi} \chi$. From (a), Success, and Adjunction we get $w \Vdash I^\phi(\phi \wedge \psi)$, thus, by (SX), $f_\phi(w) \subseteq |\phi \wedge \psi|$ and $t(\phi \wedge \psi) \leq t(\phi)$. Also, $w \Vdash I^{\phi \wedge \psi} \phi$ (from Success $\vDash I^{\phi \wedge \psi}(\phi \wedge \psi)$ and Simplification). By (SX) again, $f_{\phi \wedge \psi}(w) \subseteq |\phi|$ and (of course) $t(\phi) \leq t(\phi \wedge \psi)$. Thus, by (C3) $f_\phi(w) = f_{\phi \wedge \psi}(w)$, and $t(\phi \wedge \psi) = t(\phi)$ (by antisymmetry of content parthood). Next, from (b) and (SX) again, $f_{\phi \wedge \psi}(w) \subseteq |\chi|$ and $t(\chi) \leq t(\phi \wedge \psi)$. Therefore, $f_{\phi \wedge \psi}(w) = f_\phi(w) \subseteq |\chi|$ and $t(\chi) \leq t(\phi) = t(\phi \wedge \psi)$. Thus by (SX) again, $w \Vdash I^\phi \chi$. □

Special Transitivity has good instances. $I^\phi \psi$: as you imagine that John has won the lottery, you imagine that he has a lot of money. $I^{\phi \wedge \psi} \chi$: as you imagine that John has won the lottery and has a lot of money, you imagine that he is to pay substantive amounts of taxes. Thus, $I^\phi \chi$: as you imagine that John has won the lottery, you imagine that he is to pay substantive amounts of taxes.

Special Transitivity for the imagination operator may face counterexamples. Here's a situation suggested by Claudio Calosi, that may do. $I^\phi \psi$: given the input that I am wearing a red shirt in Pamplona, I imagine that I am being chased by bulls. $I^{\phi \wedge \psi} \chi$: given the input that I am being chased by bulls on the streets of Pamplona while wearing a red shirt, I imagine that I die on the street. But it's not the case that $I^\phi \chi$: Given that I am wearing a red shirt in Pamplona, I don't imagine that I die on its streets. So it might be that (C3) has to go for imagination, in spite of its usefulness.[3]

3.3 Hyperintensional belief revision

One automatically gets all of (C1)-(C4) if one imposes a plausibility ordering on W, thereby getting a system of spheres in the style of (Lewis, 1973). One adds to the semantics a function, $\$$, assigning to each w a finite set of nested subsets of W (the spheres): $\$(w) = \{S_0^w, S_1^w, ..., S_n^w\}$, with $n \in \mathbb{N}$, such that $S_0^w \subseteq S_1^w \subseteq ... \subseteq S_n^w = W$. Next, for each $\phi \in \mathcal{L}$ and $w \in W$, $f_\phi(w)$ goes thus: if $|\phi| = \emptyset$, then $f_\phi(w) = \emptyset$. Otherwise, $f_\phi(w) = S_i^w \cap |\phi|$, where $S_i^w \in \$(w)$ is the smallest sphere such that $S_i^w \cap |\phi| \neq \emptyset$.

In the *Erkenntnis* paper (Berto, 2018), I used this set-up to deal with AGM belief revision theory (Alchourrón, Gärdenfors, & Makinson, 1985). The feature of TSIMs under the spotlight then, is hyperintensionality. We relabel '$X^\phi \psi$' as '$B^\phi \psi$' and read it as 'Conditional on ϕ, one believes that ψ', or: 'After revising by ϕ, one believes that ψ'.

The accessibility $wR_\phi w_1$ now has us look at the *most plausible* worlds w_1 where ϕ holds, given the system of beliefs of the agent located at w, as modeled by the spheres. As in the (Grove, 1988) reformulation of the Lewisian insight, we don't demand that $w \in S_0^w$, that is, the relevant world be in the innermost sphere: in Lewis' terminology, we have a system of spheres which is not even weakly centered. That's because our spheres do not express objective world similarity, but subjective world plausibility, or belief entrenchment. The innermost sphere at the core, S_0^w, gives the most plausible worlds for the agent located at w; w itself need not be among the innermost worlds, for the agent may have false beliefs.

The relevant TSIM reduces the logical idealization of cognitive agents affecting similar operators in doxastic and epistemic logics, as well as in AGM. The first postulate for belief revision in (Alchourrón et al., 1985), (K*1), has it that $K * \phi$ (belief set K after revision by ϕ) is closed under the full strength of classical logical consequence. Postulate (K*5) trivializes belief sets revised in the light of inconsistent information: if ϕ is a logical inconsistency, then $K * \phi = K_\perp$, the trivial belief set; agents who revise

---
[3] In (Berto, 2017a), I suggested that if one resorts to an extended semantics that uses impossible worlds (of a non-adjunctive kind) besides possible ones, one can have (C3) and its welcome child, Substitutivity, without having Special Transitivity because Simplification and/or Adjunction can fail in such a framework. The proof of Special Transitivity essentially uses both, whereas the one of Substitutivity doesn't.

via inconsistent inputs trivially believe everything. And postulate (K*6) requires that, if ϕ and ψ are logically equivalent, then $K * \phi = K * \psi$, that is, revising by either gives the same belief set.

These principles are rather implausible for agents like us all. Against (K*1), our belief states need not be closed under classical logical consequence (perhaps under any kind of monotonic logical consequence: see, e.g., Jago, 2014 for extended discussion). Against (K*5), we do not trivially believe everything just because we occasionally hold inconsistent beliefs, and we should not be modeled as undergoing a trivialization of our belief system just because we can be, as we occasionally are, exposed to inconsistent information (given that information is not factive). Against (K*6), it is well known that how we revise our beliefs, as well as our preferences, is subject to what psychologists call *framing effects* (Kahneman & Tversky, 1984): logically or necessarily equivalent contents can trigger different revisions depending on how they are presented. Agents may revise their beliefs in one way when told they have 60% chances of succeeding in a task, in another way when told they have 40% chances of failing.

Our TSIM now takes care of all of these. Success (guaranteed again by (C1)) mirrors the Success postulate of AGM – After revising by ϕ, one does believe ϕ:

(**Success**) $\vDash B^\phi \phi$

(Notice that there is no problem with this holding unrestrictedly, as what we are modeling is static belief revision. Things may go differently for dynamic belief revision when, e.g., Moore formulas are concerned: see van Benthem & Smets, 2015.)

Belief revision is not automatically trivialized by incoming inconsistent information. Against principles such as AGM's (K*5), the following ensures that we do not come to believe arbitrary, irrelevant things just because we have taken on board explicitly inconsistent information:

(**Explosion**) $\nvDash B^{\phi \wedge \neg \phi} \psi$

Countermodel. Let $W = \{w\}$, $t(p) \neq t(q)$. $|p \wedge \neg p| = \emptyset$, thus $f_{p \wedge \neg p}(w) = \emptyset \leq |q|$. However, $t(q) \not\leq t(p \wedge \neg p) = t(p) \oplus t(\neg p) = t(p)$. Thus, by (S$X$), $w \nVdash B^{p \wedge \neg p} q$. □

Although there is no possible world where a contradiction is true, inconsistent information may still be about *something*. In general $\phi \wedge \neg \phi$ is not

The Theory of Topic-Sensitive Intentional Modals

contentless: its topic is whatever ϕ is about, and this may not include the topic of ψ (*Snow is white and not white* is about snow's being white, not about grass' being purple).[4]

The counterpart of AGM's (K*6) fails (where $\phi \equiv \psi$ abbreviates $\phi \prec \psi \wedge \psi \prec \phi$), thereby modelling:

(**Framing**) $\{B^\phi \chi, \phi \equiv \psi\} \nvDash B^\psi \chi$

Countermodel. Let $W = \{w, w_1\}$, $wR_p w_1$, $wR_q w_1$, $w \nVdash p$, $w \nVdash q$, $w_1 \Vdash r$, $t(p) = t(r) \neq t(q)$. Then $f_p(w) \subseteq |r|$ and $t(r) \leq t(p)$, thus by (SX), $w \Vdash B^p r$. Also, by (C1), $w_1 \Vdash p$ and $w_1 \Vdash q$, thus $|p| \subseteq |q|$ and $|q| \subseteq |p|$. Then by (S$\prec$) and (S$\wedge$), $w \Vdash p \equiv q$. But although $f_q(w) \subseteq |r|$, $t(r) \nleq t(q)$, thus $w \nVdash B^q r$. □

Again, failure of topic-preservation does the hyperintensional trick of differentiating necessarily equivalent contents. This invalidity allows a proper appreciation of framing effects: after being informed that one's probability of making it to the short list is $1/3$, one believes that one should apply for the job. But after being informed that one's probability of failing the short list is $2/3$, one does not believe that it's worth applying. There is no way that the chances of making it are $1/3$ without the chances of failing being $2/3$ and vice versa, but one has been caught into Framing.

However, by having (C3) (relabeled, for obvious reasons, as 'Principle of Equivalents in Plausibility' in the context of $B^\phi \psi$) the system of spheres allows a *limited recovery* of the idea encoded in principles like AGM's (K*6), thanks to Substitutivity – looking at it again:

(**Substitutivity**) $\{B^\phi \psi, B^\psi \phi, B^\phi \chi\} \vDash B^\psi \chi$

'Equivalents in plausibility' are now formulas ϕ and ψ such that, when we revise by either, we come to believe the other. Substitutivity now says that such equivalents can be replaced *salva veritate* as inputs for belief revision: when we revise by either, we come to have the same beliefs.[5] The

---

[4]Here's where non-classical frameworks get a revenge. Even if our TSIMs are technically not explosive, they do satisfy 'small explosion' principles like $\vDash B^{\phi \wedge \neg \phi \wedge \psi} \neg \psi$ (for, trivially, $\phi \wedge \neg \phi \wedge \psi$ is true nowhere, *and* topicality is preserved here). A framework expanded to include non-normal or impossible worlds where a contradiction can be true would help against such small detonations. I have used such a framework to model intentional operators in (Berto, 2014, 2017b).

[5]The behavior in our framework of (counterparts of) AGM principles other than (K*1),

hyperintensional belief revision operator is, overall, well-behaved qua nonmonotonic operator (it is very much unlike the KRI, in this respect). Neither Factivity nor Transitivity holds, whereas it validates Success, Restricted Transitivity, and also the following (thanks to (C3) again):

(**Cautious Monotonicity**) $\{B^\phi \psi, B^\phi \chi\} \models B^{\phi \wedge \psi} \chi$

Proof. Suppose (a) $w \Vdash B^\phi \psi$ and (b) $w \Vdash B^\phi \chi$. From (a), Success ($\models$ $B^\phi \phi$), and Adjunction, we get $w \Vdash B^\phi(\phi \wedge \psi)$, thus by (S$X$), $f_\phi(w) \subseteq |\phi \wedge \psi|$. Also, $w \Vdash B^{\phi \wedge \psi} \phi$ (from Success $\models B^{\phi \wedge \psi}(\phi \wedge \psi)$ and Simplification), so by (SX) again, $f_{\phi \wedge \psi}(w) \subseteq |\phi|$. Then, by (C3), $f_\phi(w) = f_{\phi \wedge \psi}(w)$. From (b) and (S$X$) again, we get $f_\phi(w) \subseteq |\chi|$, thus $f_{\phi \wedge \psi}(w) \subseteq |\chi|$. Also, $t(\chi) \leq t(\phi) \oplus t(\psi) = t(\phi \wedge \psi)$. Thus, $w \Vdash B^{\phi \wedge \psi} \chi$. □

Special Transitivity and Cautious Monotonicity are defended in (Board, 2004, p. 56), as 'principles of informational economy'. They correspond to the Cut and Cautious Monotonicity principles for non-monotonic logic considered in (Kraus, Lehmann, & Magidor, 1990). By satisfying these as well as Success, our $B^\phi \psi$ complies, thus, with Gabbay's (1985) minimal conditions for nonmonotonic entailments.

(K*2), (K*5) and (K*6) is somewhat less interesting, for that's not where the original features of the theory emerge. But in (Berto, 2018) I mentioned a peculiar asymmetry related to the AGM principles (K*7) and (K*8). A natural counterpart of (K*7) (see Board, 2004, p. 55) fails in our semantics:

$$\{\neg B^\phi \neg \psi, B^{\phi \wedge \psi} \chi\} \not\models B^\phi(\psi \prec \chi)$$

Countermodel. Let $W = \{w\}, f_p(w) = \emptyset, f_{p \wedge q} = \emptyset, t(p) \neq t(q) = t(r)$. Then by (S$X$), $w \not\Vdash B^p \neg q$ because $t(\neg q) = t(q) \not\leq t(p)$, so $w \Vdash \neg B^p \neg q$; and $w \Vdash B^{p \wedge q} r$, because (trivially) $f_{p \wedge q}(w) \subseteq |r|$, and $t(r) = t(q) \leq t(p) \oplus t(q) = t(p \wedge q)$. However, $w \not\Vdash B^p(q \prec r)$, because $t(q \prec r) = t(q) \oplus t(r) = t(q) \not\leq t(p)$. □

On the other hand, a natural counterpart of (K*8) (see Board, 2004, p. 55), obtained by flipping premise and conclusion in the former, holds:

$$\{\neg B^\phi \neg \psi, B^\phi(\psi \prec \chi)\} \models B^{\phi \wedge \psi} \chi$$

Proof. Suppose (a) $w \Vdash \neg B^\phi \neg \psi$ and (b) $w \Vdash B^\phi(\psi \prec \chi)$. By (a) and (S¬), $w \not\Vdash B^\phi \neg \psi$, that is: either $f_\phi(w) \not\subseteq |\neg \psi|$, that is, $f_\phi(w) \cap |\psi| \neq \emptyset$, or $t(\neg \psi) = t(\psi) \not\leq t(\phi)$. But it can't be the latter, because by (b) and (SX), $t(\psi \prec \chi) = t(\psi) \oplus t(\chi) \leq t(\phi)$, thus in particular $t(\psi) \leq t(\phi)$; so it must be the former. Applying Condition (C4) to it, $f_{\phi \wedge \psi}(w) \subseteq f_\phi(w)$. By (C1), $f_{\phi \wedge \psi}(w) \subseteq |\phi \wedge \psi|$, so by (S$\wedge$), $f_{\phi \wedge \psi}(w) \subseteq |\psi|$. By (b) and (S$X$) again, $f_\phi(w) \subseteq |\psi \prec \chi|$. Putting things together: $f_{\phi \wedge \psi}(w) \subseteq f_\phi(w) \subseteq |\psi \prec \chi|$, so $f_{\phi \wedge \psi}(w) \subseteq |\psi \prec \chi|$; and since $f_{\phi \wedge \psi}(w) \subseteq |\psi|$, then by modus ponens $f_{\phi \wedge \psi}(w) \subseteq |\chi|$. Also, by (b) again, $t(\psi) \oplus t(\chi) \leq t(\phi) \leq t(\phi \wedge \psi)$, thus $t(\chi) \leq t(\phi \wedge \psi)$. Thus, by (S$X$), $w \Vdash B^{\phi \wedge \psi} \chi$. □

4 Further work

Some work in TSIM theory, beyond what I have summarized in this paper, is already being carried out. Sound and complete axiomatizations of the semantics proposed above are being developed by Alessandro Giordani (forthcoming) and Aybüke Ozgün. Heinrich Wansing is working on how to combine the semantics of imagination from Section 3.2 above with his agentive STIT logic of imagination (Olkhovikov & Wansing, 2018; Wansing, 2015). And there are also developments in the direction of dynamic epistemic and doxastic logic (Ozgün again, and Peter Hawke): the hyperintensional belief revision operator is static, but we are exploring some ideas on how we make the framework dynamic.

Further possible areas of research include, e.g., moving to a first-order language, a nice question then being how we want topicality to work there. Perhaps the biggest open issue concerning topicality is the following. All the ways of playing with the TSIMs explored so far tamper only with accessibilities. None tampers with their topic-sensitivity. All our $X^\phi\psi$'s embed a rather draconian topicality or aboutness constraint: ψ must be fully on-topic with respect to ϕ, or what ψ is about must be fully included in what ϕ is about. I haven't explored how to play with the mereology of topics yet, but there are reasons to relax such a constraint, allowing, e.g., partial overlap of topics rather than full inclusion, for various purposes. If we allow ψ to only be partly on topic with respect to ϕ, this brings '$X_\phi\psi$' in the vicinity of a variably strict relevant conditional. And there are surely other options for more complicated topic-embeddings. It's a nice territory and I hope more people get interested in exploring it.

References

Alchourrón, C., Gärdenfors, P., & Makinson, D. (1985). On the logic of theory change: Partial meet functions for contraction and revision. *Journal of Symbolic Logic*, *50*, 510–30.

Angell, R. (1977). Three systems of first degree entailment. *Journal of Symbolic Logic*, *47*, 147.

Asheim, G., & Sövik, Y. (2005). Preference-based belief operators. *Mathematical Social Sciences*, *50*, 61–82.

Baltag, A., & Smets, S. (2008). A qualitative theory of dynamic interactive belief revision. In G. Bonanno, W. van der Hoek, & M. Wooldridge (Eds.), *Logic and the Foundations of Game and Decision Theory* (pp. 9–58). Amsterdam: Amsterdam University Press.

Berto, F. (2014). On conceiving the inconsistent. *Proceedings of the Aristotelian Society, 114*, 21–103.

Berto, F. (2017a). Aboutness in imagination. *Philosophical Studies, 175*, 1871–86.

Berto, F. (2017b). Impossible worlds and the logic of imagination. *Erkenntnis, 82*, 1277–97.

Berto, F. (2018). Simple hyperintensional belief revision. *Erkenntnis*.

Berto, F., & Hawke, P. (2018). Knowability relative to information. *Mind*.

Board, O. (2004). Dynamic interactive epistemology. *Games and Economic Behaviour, 49*, 49–80.

Byrne, R. (2005). *The Rational Imagination. How People Create Alternatives to Reality*. Cambridge, Mass.: MIT Press.

Correia, F. (2004). Semantics for analytic containment. *Studia Logica, 77*, 87–104.

Crane, T. (2013). *The Objects of Thought*. Oxford: Oxford University Press.

Dretske, F. (1970). Epistemic operators. *Journal of Philosophy, 67*(24), 1007–1023.

Dretske, F. (1999). *Knowledge and the Flow of Information*. CLSI Publications.

Fagin, R., & Halpern, J. (1988). Belief, awareness and limited reasoning. *Artificial Intelligence, 34*, 39–76.

Ferguson, T. (2014). A computational interpretation of conceptivism. *Journal of Applied Non-Classical Logics, 24*(4), 333–367.

Fine, K. (1986). Analytic implication. *Notre Dame Journal of Formal Logic, 27*, 169–79.

Fine, K. (2014). Truthmaker semantics for intuitionistic logic. *Journal of Philosophical Logic, 43*, 549–77.

Fine, K. (2016). Angellic content. *Journal of Philosophical Logic, 45*(2), 199–226.

Floridi, L. (2015). Semantic conceptions of information. In E. N. Zalta (Ed.), *The Stanford Encyclopedia of Philosophy* (Spring 2015 ed.). Metaphysics Research Lab, Stanford University. https://plato.stanford.edu/archives/spr2015/entries/information-semantic.

Gabbay, D. (1985). *Theoretical Foundations for Non-Monotonic Reasoning.* Berlin: Springer.

Gendler, T. (2000). The puzzle of imaginative resistance. *Journal of Philosophy, 97,* 55–81.

Girard, P., & Rott, H. (2014). Belief revision and dynamic logic. In A. Baltag & S. Smets (Eds.), *Johan van Benthem on Logic and Information Dynamics* (pp. 203–33). Dordrecht: Springer.

Grove, A. (1988). Two modellings for theory change. *Journal of Philosophical Logic, 17,* 157–170.

Harman, G. (1973). *Thought.* Princeton University Press.

Hawke, P. (2017). Theories of aboutness. *Australasian Journal of Philosophy,* 1–27.

Hintikka, J. (1962). *Knowledge and Belief. An Introduction to the Logic of the Two Notions.* Ithaca, NY: Cornell University Press.

Jago, M. (2014). *The Impossible. An Essay on Hyperintensionality.* Oxford: Oxford University Press.

Kahneman, D., & Tversky, A. (1984). Choices, values, and frames. *American Psychologist, 39,* 341–50.

Kind, A. (2016). Imagining under constraints. In A. Kind & P. Kung (Eds.), *Knowledge Through Imagination* (pp. 145–59). Oxford: Oxford University Press.

Kind, A., & Kung, P. (Eds.). (2016). *Knowledge through Imagination.* Oxford: Oxford University Press.

Kraus, S., Lehmann, D., & Magidor, M. (1990). Nonmonotonic reasoning, preferential models and cumulative logics. *Artificial Intelligence, 44,* 167–207.

Kripke, S. (2011a). Nozick on knowledge. In *Philosophical Troubles: Collected Papers, Volume 1.* Oxford University Press.

Kripke, S. (2011b). On two paradoxes of knowledge. In *Philosophical Troubles: Collected Papers, Volume 1.* Oxford University Press.

Langland-Hassan, P. (2016). On choosing what to imagine. In A. Kind & P. Kung (Eds.), *Knowledge Through Imagination* (pp. 61–84). Oxford: Oxford University Press.

Leitgeb, H., & Segerberg, K. (2005). Dynamic doxastic logic: Why, how, and where to? *Synthèse, 155,* 167–90.

Lewis, D. (1973). *Counterfactuals.* Oxford: Blackwell.

Lindström, S., & Rabinowicz, W. (1999). DDL unlimited: Dynamic doxastic logic for introspective agents. *Erkenntnis, 50,* 353–85.

Markman, K., Klein, W., & Surh, J. (Eds.). (2009). *Handbook of Imagination and Mental Simulation*. New York: Taylor and Francis.

Nichols, S., & Stich, S. (2003). *Mindreading. An Integrated Account of Pretence, Self-Awareness, and Understanding Other Minds*. Oxford: Oxford University Press.

Nozick, R. (1981). *Philosophical Explanations*. Harvard University Press.

Olkhovikov, G., & Wansing, H. (2018). An axiomatic system and a tableau calculus for STIT imagination logic. *Journal of Philosophical Logic*, 47(2), 259–79.

Parry, W. (1933). Ein axiomensystem für eine neue art von implikation (analitische implikation). *Ergebnisse eines Mathematischen Kolloquiums*, 4, 5–6.

Schipper, B. (2015). Awareness. In H. van Ditmarsch, J. Halpern, W. van der Hoek, & B. Kooi (Eds.), *Handbook of Epistemic Logic* (pp. 79–146). London: College Publications.

Segerberg, K. (1995). Belief revision from the point of view of doxastic logic. *Bulletin of the IGPL*, 3, 535–53.

Sharon, A., & Spectre, L. (2010). Dogmatism repuzzled. *Philosophical Studies*, 148(2), 307–321.

Sharon, A., & Spectre, L. (2017). Evidence and the openess of knowledge. *Philosophical Studies*, 174, 1001–1037.

Spohn, W. (1988). Ordinal conditional functions: a dynamic theory of epistemic states. In L. Hrper & B. Skyrms (Eds.), *Causation in Decision, Belief Change, and Statistics* (Vol. 2, pp. 105–34). Dordrecht: Kluwer.

van Benthem, J. (2007). Dynamic logic for belief revision. *Journal of Applied Non-Classical Logic*, 17, 129–55.

van Benthem, J. (2011). *Logical Dynamics of Information and Interaction*. Cambridge: Cambridge University Press.

van Benthem, J., & Smets, S. (2015). Dynamic logics of belief change. In H. van Ditmarsch, J. Halpern, W. van der Hoek, & B. Kooi (Eds.), *Handbook of Epistemic Logic* (pp. 313–93). London: College Publications.

van Ditmarsch, H. (2005). Prolegomena to dynamic logic for belief revision. *Synthèse*, 147, 229–75.

van Ditmarsch, H., van der Hoek, W., & Kooi, B. (2008). *Dynamic Epistemic Logic*. Dordrecht: Springer.

van Fraassen, B. (1969). Facts and tautological entailments. *Journal of Philosophy*, 66, 477–87.

Van Leeuwen, N. (2016). The imaginative agent. In A. Kind & P. Kung (Eds.), *Knowledge Through Imagination* (pp. 85–111). Oxford: Oxford University Press.

Wansing, H. (2015). Remarks on the logic of imagination. a step towards understanding doxastic control through imagination. *Synthese, On line first*.

Williamson, T. (2007). *The Philosophy of Philosophy*. Oxford: Blackwell.

Williamson, T. (2016). Knowing by imagining. In A. Kind & P. Kung (Eds.), *Knowledge Through Imagination* (pp. 113–23). Oxford: Oxford University Press.

Yablo, S. (2014). *Aboutness*. Princeton: Princeton University Press.

Francesco Berto
University of St Andrews and University of Amsterdam
United Kingdom and The Netherlands
E-mail: `fb96@st-andrews.ac.uk`

Categoricity and Negation.
A Note on Kripke's Affirmativism

CONSTANTIN C. BRÎNCUŞ AND IULIAN D. TOADER

Abstract: The idea that an adequate language for science needs a negation operator was recently dismissed by Kripke as "yet another dogma of empiricism". That a scientist could, and even should, drop negation implies at least three points: 1. negativist theories, i.e., theories formulated in languages that include negation, are conservative extensions of their affirmativist versions; 2. negativist theories have no serious advantages over their affirmativist versions; 3. negativist theories are dispensable and should better be replaced by their affirmativist versions. We argue that all three points are problematic.

Keywords: negation, categoricity, idealization

1 Introduction

Kripke (2015) argues that an affirmativist language is adequate for science: "In strictly scientific discourse, or serious discourse generally, limning the true and ultimate nature of reality, restriction to affirmativist terminology is the way to go" (p. 384). So if you're a scientist you could, and even should, drop negation. This implies that, in science, negativist theories are conservative extensions of their affirmativist versions, have no serious advantages over their affirmativist versions, and are actually dispensable and so better replaced by their their affirmativist versions. But we think that all these three points are problematic. We will take them in turn.[1]

We point out, first, that negativist theories can be conservative extensions of their affirmativist version provided that one takes the notion of logical complement as an affirmativist notion. Secondly, we describe what Kripke

---

[1] We should note that, in an *Addendum* to his paper, Kripke confessed that it is rather a "parody" of some of Quine's own arguments, but also that he cannot deny that his argument is sound, since he would thereby be lapsing into a negativistic idiom. Parody or not, suppose that we take Kripke's paper seriously. Then our discussion below entails that one should either reject his argument or dismiss this supposition.

took the practical advantages of affirmativist theories to be, and then consider the view that an epistemic advantage of negativist theories is that they are simpler than their affirmativist versions: the simplifying effects of classical negation are analogous to those of imaginary numbers in mathematics. But we emphasize that while in mathematics such effects might be explained in terms of categoricity, classical logic is non-categorical, for it admits of non-normal interpretations. We also prove that eliminating negation does not help one get rid of such interpretations: positive (and affirmativist) logic is non-categorical as well. Finally, we argue that affirmativist restrictions on science are at odds with the typical understanding of scientific idealization.

2 Is T_- a conservative extension of T_+?

Let's start with Kripke's argument for the claim that negativist theories are conservative extensions of their affirmativist versions. Consider the affirmativist first order language, L_+, of an affirmativist theory, T_+. L_+ contains a finite list of primitive predicates, conjunction and disjunction as primitive connectives, and the universal and existential quantifiers. Is L_+ adequate for science?

Let us assume that L_+ is adequate only if negation is added. By De Morgan Laws, every sentence in the augmented language, L_-, is logically equivalent to a sentence with negation applied only to atomic formulae. Let us eliminate negation and extend L_+ by adding to each predicate P_i a predicate P_i^* for its complement, and likewise for any atomic sentence. Thus, everything expressible in L_- is expressible in the affirmativist language thus supplemented, L_+^*.

Kripke admits that the notion of complement is negativistic, but claims that this is not a problem since the argument just given is directed at negativists, not affirmativists. This argumentative move is adopted by Kripke from Quine (1960), where mentalist terms are employed in explaining to mentalists the physicalist view about the mind. Using negativist notions in Kripke's affirmativist argument directed at negativists is analogous to using mentalist notions in Quine's physicalist argument directed at mentalists. Be that as it may, if the notion of complement is negativistic, then how can L_+^* be counted as an affirmativist language? Kripke's argument for the idea that negativistic theories are conservative extensions of their affirmativist versions goes through only if one assumes that the notion of complement is an affirmativist one. But in many logical systems, especially algebraic logical

A Note on Kripke's Affirmativism

systems, the complement is precisely the notion that expresses negation.[2]

We should further note that, unlike classical positive logic, Kripke's affirmativist logic does not have a classical semantics: it eliminates not only the negation operator, but also the notion of "falsity". A conjunction $\ulcorner A \wedge B \urcorner$ is true if and only if A is true and B is true, and a disjunction $\ulcorner A \vee B \urcorner$ is true if and only if A is true or B is true. However, if the normal truth tables (NTTs) for conjunction and disjunction contain only the first line[3], then we cannot really distinguish conjunction from disjunction, since only the other three lines of the NTTs allow us to make a distinction between conjunction and disjunction. Why would "and" and "or" mean different things for the affirmativist?

But even if the affirmativist could distinguish between "and" and "or", the class of logical rules of inference and of logical truths of affirmativist logic is drastically affected. For instance, $\ulcorner A \rightarrow (B \rightarrow A) \urcorner$ could no longer be treated as a logical truth, since the material conditional is defined only on the first line of the NTT and thus, without knowing the value of B, we may be reluctant to accept $\ulcorner B \rightarrow A \urcorner$, and thus $\ulcorner A \rightarrow (B \rightarrow A) \urcorner$.

3   Negation as an ideal element

What might be the advantages of affirmativist theories? Kripke claims that affirmativism is more advantageous than negativism, because it improves the civility of our debates, for example. If you thought that by eliminating negation, you would not be able to say anymore that the negation of a true statement is false, Kripke would say that you should never state "Your opinion is false!" in the first place. Instead, what you should say is "I am reluctant to accept your view". He thinks that this may lead, in the long run, to world peace. No argument is given for this claim, other than the observation that many conflicts have been preceded by negativistic characterizations of one's opponents' assertions as "false". Kripke also thinks that affirmativism offers a solution to paradoxes, since he thinks that all of them are easily seen

---

[2] As a matter of logical fact, the negation operation could not be defined with the help of other sentential operations. This is why classical positive logic, i.e., classical logic without the negation operator but with classical semantics, is incomplete – logical truths such as Peirce's Law cannot be derived without negation.

[3] From Kripke's remark on this matter, it is natural to suppose so. He explicitly states that we do not need the third extra lines from the truth-table for conjunction, the first line is sufficient. This would suggest that the lines that contain the sign for falsity should be dismissed. If Kripke introduced a sign for the attitude of reluctance to replace the sign for falsity, then, of course, the propositional operators could be easily defined and distinguished one from another.

to invoke a negativistic notion. More precisely, without negation, paradoxes would just "disappear". However, it is not clear that this is the case. As Skolem (1952) proved, Dedekind's naive set theory still leads to inconsistency via the comprehension axiom, even if the logic used in the background is positive predicate calculus (i.e., first-order logic without negation; see also Hilbert & Bernays, 1934/1968).

Furthermore, imposing affirmativist restrictions in science assumes that the alleged practical advantages of affirmativism (e.g., less conflicts, peace, etc.) outweigh any other advantages negativist theories might have. But there is a long tradition in the philosophy of logic, stemming from Hilbert's school, which considers negation as an idealization: "Negation plays the role of an ideal element whose introduction aims at rounding off the logical system to a totality with a simpler structure, just as the system of real numbers is extended to a more perspicuous totality by the introduction of imaginary numbers." (Bernays, 1927). Negativist theories, it is claimed here, are simpler than their affirmativist versions. For example, the simplifying effects of classical negation are comparable to the simplifying effects of the imaginaries, and so eliminating negation is comparable to the elimination of the imaginaries from mathematics. Extending a logical system by adding connectives, like negation, leads to a totality with a simpler structure. As Hilbert also emphasized, negation makes possible the logical closure and completeness of a system (see, e.g., Hilbert, 1931).

This view raises some important questions. Let's assume that extending a number system by adding new objects leads to a more perspicuous totality. In particular, adding imaginary numbers to the reals forms the algebraic closure of the real numbers, but what makes an algebraic closure a more perspicuous totality? One answer to this question might be given in semantic terms like categoricity: the algebraic closure of the real number field is unique up to isomorphism (Steinitz, 1910). But what makes a closed and complete logical system a totality with a simpler structure? Could a similar answer be given in terms of categoricity?

4 The non-categoricity of classical logic

Let a semantic property of an expression be fully formalized by a calculus if and only if the expression possesses that property in every interpretation for which the calculus is sound. As is well known, however, classical logic allows for non-standard models, i.e., interpretations for which the standard

A Note on Kripke's Affirmativism

calculi remain sound and complete, but in which the logical constants have different meanings than the standard ones (Carnap, 1943). The existence of these interpretations shows that the standard propositional and quantificational calculi do not fully formalize all the semantic properties of the logical terms and, thus, fail in uniquely determining their meanings. The rules for negation, disjunction, material implication, and the quantifiers, in contrast to the rules for conjunction, do not determine all the properties of these operators as defined by their normal truth tables (NTT) and the standard semantics for quantifiers.

In propositional logic, non-normal interpretations are possible because the usual formalizations of classical logic state conditions only for logical derivability (C-implicate, in Carnap's terms) and logical theoremhood (C-truth). Thus, they can formalize only those semantic properties definable on the basis of logical consequence (L-implicate) and logical truth (L-truth). However, the semantic properties of L-exclusive and L-disjunct are not definable on this basis, thus, they are not formalized by the usual systems. Two sentences are L-exclusive if and only if they are not both true (thus, at least one is false), and two sentences are L-disjunct if and only if at least one of them is true (thus, they are not both false). Since L-exclusive and L-disjunct are not fully formalized, the principles of non-contradiction and excluded middle are not represented in the usual formalizations of classical logic.

There are two mutually exclusive types of non-normal interpretations for classical propositional operators: (I) all sentences are true, and (II) at least one sentence is false. Thus, there are non-normal interpretations in which a sentence and its negation are both true, and non-normal interpretations in which they are both false (and so, their disjunction is true and their implication is false). In type (I) interpretations, the principle of non-contradiction is violated; in type (II), the principle of excluded middle is violated. Thus, these two principles are not fully formalized, since they hold in some interpretations (i.e., in the normal interpretations), but they do not hold in others (i.e., in the non-normal interpretations). In addition, even if all propositional operators had only normal interpretations, there would still exist non-normal interpretations of the quantifiers (as shown in Carnap 1943, chapter F). In particular, there are sound interpretations of quantificational logic in which "$(\forall x)Fx$" could be interpreted as "every individual is F, and b is G", where "b" is an individual constant. Likewise, "$(\exists x)Fx$" could be interpreted as "at least one individual is F, or b is G". The possibility of these non-normal interpretations arises because, in the standard formalizations of first-order logic, a universal sentence is not deductively equivalent (C-equivalent, in

Carnap's terms) to the conjunction of all the instances of the operand, and an existential sentence is not C-equivalent with the disjunction of all the instances of the operand. The existence of such interpretations shows that the logical calculus fails in uniquely determining the meaning of logical terms. Unlike the rules for conjunction, those for negation (as well as those for disjunction, implication, and quantifiers) do not determine all the properties of these operators as defined by NTTs and standard semantics.

Kripke took Carnap (1943) as an illustration of a negativistic theory, as it includes not only rules of theoremhood, but also rules of rejection, and he thought that discussing it in more detail would be "superfluous". But we think that it is important to note that Carnap proved that if negation has a standard interpretation, then all other operators have a standard interpretation. Thus, the categoricity of negativist theories like classical logic would require the elimination of non-standard interpretations of negation. The affirmativist could point out that if one eliminates negation, its non-standard interpretations are thereby eliminated as well. This seems natural enough since the principles of non-contradiction and excluded middle do not belong to positive logic. This would suggest a shorter, if more radical, route to categoricity.

However, this assumes that a positive version of classical logic does not allow for non-normal interpretations. A closer investigation, however, shows that the elimination of negation does not immediately entail that positive logic admits of no non-normal interpretations. The normality of negation constrains disjunction (and, thus, all the other operators) to be normal only if negation is, of course, part of the system. If negation has a standard interpretation, then disjunction has a standard interpretation. This is due to the Disjunctive Syllogism (DS): $A \vee B, \neg A \vdash B$. If A and B would be false, and negation is standard (thus, $\ulcorner \neg A \urcorner$ is true), then $\ulcorner A \vee B \urcorner$ cannot be true, otherwise the rule would be unsound. However, we do not have the DS in positive logic. In the absence of negation, what happens to disjunction? As we shall see in a moment, in positive logic, A and B can be false, but $\ulcorner A \vee B \urcorner$ true.

5 The non-categoricity of positive logic

The semantic property of disjunction displayed on line D4 of its NTT is not determined by the deduction rules for disjunction. Let us analyze what happens when D4 is violated, that is, what kind of non-normal interpretation

A Note on Kripke's Affirmativism

would result from this violation, and then let us check whether this interpretation is non-empty. Let us assume that D4 is violated with respect to the propositional constants A and B. On this assumption, we can reason as follows:

a) A is false, B is false, and $\ulcorner A \vee B \urcorner$ is true. (by assumption)
b) A is different from B (and conversely). Proof: If A were B, then A, which is derivable from $\ulcorner A \vee A \urcorner$, would be derivable from $\ulcorner A \vee B \urcorner$ and, thus, true. But A is false. Thus, A is different from B.
c) Any sentence derivable both from A and from B is true. Proof: If a sentence is derivable both from A and from B, then it is derivable from $\ulcorner A \vee B \urcorner$, and, thus, true.
d) A does not follow from B, nor B from A. Proof: If A were derivable from B, since it is derivable from itself, it would be derivable from $A \vee B$ and, thus, true. But A is false.
e) $\ulcorner A \rightarrow B \urcorner$ is false. Proof: $A \vee B \vdash ((A \rightarrow B) \rightarrow B)$. Thus $\ulcorner (A \rightarrow B) \rightarrow B \urcorner$ is true. Since B is false, then according to line I4 in the NTT, $\ulcorner A \rightarrow B \urcorner$ has to be false. But since A is false as well, then I4 is violated with respect to A and B.

Thus, we can see that the violation of D4 would lead to a non-normal interpretation of the positive calculus in which the truth value of $\ulcorner A \rightarrow B \urcorner$ on line I4 in the NTT is not determined by the rules of material implication, since this operator is non-extensional, i.e., it behaves normally in its main occurrence in $\ulcorner (A \rightarrow B) \rightarrow B \urcorner$, but non-normally in $\ulcorner A \rightarrow B \urcorner$. What we have to examine now is whether the non-normal interpretation with the features just sketched is non-empty. That, indeed, it is non-empty will be shown by the construction of an example, namely, the construction of an interpretation, V_+, as follows:

1. if p is a theorem of the positive calculus, then $V_+(p)$ is true.
2. if p is not a theorem of the positive calculus, then $V_+(p)$ is false, but in the following two cases:
   (a) $V_+(A \vee B)$ is true.
   (b) For every C, if $A \vee B \vdash C$, then $V_+(C)$ is true.

This interpretation assigns 'truth' to every theorem of positive calculus and 'false' to every non-theorem of the positive calculus, except in the two cases described: namely, it will assign 'truth' to $\ulcorner A \vee B \urcorner$ and to all formulas derivable from it. Thus, since it assigns a determinate truth value to all formulas of positive logic, V_+ is a full interpretation of the positive calculus.

What we have to show now is that the positive calculus remains sound under this interpretation.

Proof: Let Γ be a set of premises and σ an arbitrary sentence in the language of positive calculus, and let us further suppose that $\Gamma \vdash \sigma$. There are three cases to be considered:
1. if in Γ we have only theorems, then σ will be a theorem and, thus, true in V_+.
2. if in Γ we have a non-theorem different from $\ulcorner A \vee B \urcorner$ and from any C derivable from $\ulcorner A \vee B \urcorner$, then the sequent $\Gamma \vdash \sigma$ will be valid even if σ is false.
3. if Γ contains a set of theorems (Δ) and, in addition, it contains a non-theorem which is either $\ulcorner A \vee B \urcorner$, or any C derivable from $\ulcorner A \vee B \urcorner$, then σ either follows from Δ, and thus it is true, or it follows from $\ulcorner A \vee B \urcorner$, and thus it is also true.

Therefore, the calculus of positive logic remains sound in the interpretation V_+. However, as constructed, this interpretation is non-normal, since it makes a disjunction true, although both of its disjuncts are false. What this shows is that the existence of non-normal interpretations of a logical calculus does not require the non-normality of logical negation. Even without negation, the rules for disjunction do not completely determine the semantic properties of disjunction as defined by its NTT. In addition, since positive quantificational logic is obtained by adding the standard rules for the existential and universal quantifiers to the positive fragment of classical logic, it is in no different position than classical quantificational logic with respect to non-normality: the non-normal interpretations of the quantifiers in classical logic are also present in the case of positive quantificational logic.

Furthermore, things are similar in affirmativist logic, because the elimination rule for disjunction involves no negation, and it is responsible for the non-standard models for disjunction. More exactly, in affirmativist logic, one may take a disjunction to be true, although one may be reluctant to accept its disjuncts. Therefore, the elimination of negation and falsity does not entail that there are no non-standard interpretations.

To take stock, we have argued that negativist theories may be conservative extensions of their affirmativist versions, but only if the notion of logical complement is accepted as an affirmativist notion, and that these affirmativist versions may have important practical advantages in that they may be less conflictual and more civil, but that one should not overlook some important epistemic advantages of negativist theories, e.g., simplicity. We have also pointed out that negativist theories do not obstruct purported semantic advantages of their affirmativist versions: since positive and affirmativist logics are non-categorical, one cannot maintain that categoricity is lost due to the addition of the negation operator. Thus, the affirmativist's elimination of negation cannot be justified on such semantic grounds.

6 Affirmativism and idealization

Let's further suppose that one has strong enough reasons to prefer practical to epistemic advantages, and so let's suppose that one accepts the affirmativist restrictions on science. Consider again imaginary numbers and their simplifying effects. The latter could also be explained in terms of linear factorization, i.e., via the fact that the only irreducible polynomials are those of degree one. This is stated by the Fundamental Theorem of Algebra: "every equation of degree n has n roots". The theorem is, of course, false in the real number system, but true in its algebraic closure, i.e., in the complex number system. According to the affirmativist, however, one should not say that the theorem is false in the real number system, since falsity has been eliminated, but only that one is reluctant to accept the theorem in the real number system. Analogously, one should not say that, for instance, the ideal gas law is false for real gases, but only that one is reluctant to accept it. More generally, one should not say that an idealized statement is false for non-idealized systems, but only that one is reluctant to accept that statement. However, this is entirely missing the point of scientific idealization. In science, idealized statements are typically rather unreluctantly accepted, for even though they are false for non-idealized systems, they are thought to have great explanatory power (Toader, 2015).

In conclusion, *pace* Kripke, we argued that an affirmativist first order language, L_+, cannot be adequate for science. We believe that if Kripke's affirmativist view is taken seriously, then our argument entails that his view should be rejected. More precisely, our argument entails that L_+ cannot be taken as the object language of a genuine scientific theory. But this leaves other questions open. Could L_+ be adequate as a metalanguage for science, i.e., as the language of the metatheory of a scientific theory? In response to this, we would like just to note here a recent effort to do model theory in the framework of positive logic: "a non first order analogue of classical model theory where compactness is kept at the expense of negation" (Ben-Yaacov, 2003). Such an approach has been taken, for instance, to the formal semantics of quantum mechanics (see, e.g., Zilber, 2016). But we are fine, for the time being, with a negationless metatheory.

References

Ben-Yaacov, I. (2003). Positive model theory and compact abstract theories. *Journal of Mathematical Logic*, *3*, 85–118.

Bernays, P. (1927). Problems of theoretical logic. *Lecture given at the 56th meeting of German philologists and teachers in Göttingen in September 1927.*

Carnap, R. (1943). *Formalization of Logic*. Cambridge, Mass.: Harvard University Press.

Hilbert, D. (1931). The grounding of elementary number theory. *Mathematische Annalen*, *104*, 485–494.

Hilbert, D., & Bernays, P. (1934/1968). *Grundlagen der Mathematik, I*. Berlin, Heidelberg, New York: Springer.

Kripke, S. (2015). Yet another dogma of logical empiricism. *Philosophy and Phenomenological Research*, *91*, 381–385.

Quine, W. V. O. (1960). *Word and Object*. Cambridge, MIT Press.

Skolem, T. (1952). A remark on a set theory based on positive logic. *Det Kongelige Norske Videnskabers Selskab, Forhandlinger*, *25*, 112–116.

Steinitz, E. (1910). *Algebraische Theorie der Körper*. Journal für Mathematik, vol. 137.

Toader, I. D. (2015). Against harmony: Infinite idealizations and causal explanation. *Boston Studies in the Philosophy and History of Science*, *313*, 291–301.

Zilber, B. (2016). The geometric semantics of algebraic quantum mechanics. *Available at: https://arxiv.org/abs/1604.07745.*

Constantin C. Brîncuş
University of Bucharest
Romania
E-mail: c.brincus@yahoo.com

Iulian D. Toader
University of Salzburg
Austria
E-mail: Iulian.Toader@sbg.ac.at

A Justification Logic for Argument Evaluation

ALFREDO BURRIEZA AND ANTONIO YUSTE-GINEL[1]

Abstract: The central aim of this paper consists in offering a formal analysis of the notion of preference between deductive arguments for supporting a given claim. After introducing the problem and its motivations, we propose a new development of justification logic that combines several existing tools of the literature and show how a list of argumentative and epistemic notions are expressible in its language. Later on, we discuss how these notions should be related to the notion of preference and explain how to represent this relation in the proposed framework.

Keywords: argument evaluation, justification logic, doxastic logic

1 Introduction

A recent branch of studies within the field of epistemic logic has focused in the analysis of the justifications that an epistemic agent possesses in order to support one belief or another regarding a specific topic. This branch is being developed using different formal tools. Some authors have captured the justification component by the use of justification logic (Artemov, 2008; Artemov & Fitting, 2016; Artemov & Nogina, 2005; Baltag, Renne, & Smets, 2012, 2014). Following this option, some of the epistemic effects caused by dynamic processes involving justifications, such as evidence elimination caused by communications, have also been studied (Renne, 2012). Some others have used neighbourhood semantics with the same purpose (van Benthem, Fernández-Duque, & Pacuit, 2012, 2014; van Benthem & Pacuit, 2011). Besides, in the very recent years, approaches mixing classic epistemic logic with abstract argumentation frameworks have appeared (Grossi & van der Hoek, 2014; Shi, Smets, & Velázquez-Quesada, 2017, 2018).

Nevertheless, there is something that has not received enough attention yet: how epistemic agents might prefer certain justifications to others, in

---

[1] This research is partially supported by MECD-FPU 2016/04113 and TIN2015-72709-EXP

order to have better pieces of evidence for achieving a given goal. It seems clear that the manner agents should prefer one justification to another depends on a variety of factors. Therefore, the notion of preference between justifications should be sensitive to several contextual features, such as the goal of evaluation, the kinds of justification that are being assessed, the epistemic perspective of agents, etc. In this paper, we aim to develop a formal analysis of the notion of preference between a particular kind of justifications: *deductive arguments*. Assuming the context dependence of preferences we have just mentioned, another contextual assumption is made: we will focus on situations in which the goal of argument evaluation is supporting a given claim.

Therefore, the central objective of this paper can be expressed as follows. Let a be an epistemic agent, let φ be a claim and let t, s be a couple of (possibly different) deductive arguments:

(i) Which of the arguments (t or s) should be preferred by a if she intends to support φ?

The motivation for this question and the potential applications of our answer are diverse and belong to different fields. For example, knowing which argument is better for a given agent would imply knowing which argument will be selected by her in the context of an argumentative dialogue when she tries to support a given thesis. It would also offer an approach to face the problem of how an agent establishes her beliefs solidly enough to convert them into knowledge. Regarding formal argumentation theory, an appropriate answer to *(i)* represents a formal theory of deductive argument evaluation.

This paper is organized as follows. In section 2 we introduce a logical framework mixing some existing tools of justification logic and doxastic logic in which a set of argumentative and doxastic notions are expressible. In section 3 we develop an analysis of the notion of preference between deductive arguments for supporting a given claim. For that purpose, we first discuss some intuitive principles that an appropriate notion should satisfy and then we define an operator using the object language of the new logic that indeed satisfies those principles. In section 4 we briefly sketch some conclusions and mention some possible paths for future work.

A Justification Logic for Argument Evaluation

2 The logical apparatus

The general framework we are going to work with is justification logic. The main reason to choose this tool for representing arguments is a simple but appealing one: it seems that the relations between justification logic and argumentation theory have not been explored enough yet in the literature. This section is devoted to the presentation of a new logic, called AE.

2.1 Syntax

Definition 1 (Language) *Let Φ be a denumerable set of atoms and let $\mathcal{A}$ be a non empty and finite set of agents. The language $\mathcal{L}^{AE}$ is defined as the pair $\langle \mathcal{F}, \mathcal{T} \rangle$, where $\mathcal{F}$ (formulas) and $\mathcal{T}$ (terms) are defined by the following grammar:*

$$\varphi ::= p \mid \neg\varphi \mid (\varphi \to \varphi) \mid B_a\varphi \mid D_{ab}\varphi \mid t \gg \varphi \mid Com^{\leq}(t,t)$$

where $p \in \Phi$, $a, b \in \mathcal{A}$, $t \in \mathcal{T}$ and

$$t ::= c_\varphi \mid [t + t] \mid [t \cdot t] \quad \text{where } \varphi \in \mathcal{F}$$

The concepts of subformulas of a given formula $sub(\varphi)$ and subterms of a given term $sub(t)$ are defined as expected, assuming that $sub(t \gg \varphi) = \{t \gg \varphi\}$ and $sub(Com^{\leq}(t,s)) := \{Com^{\leq}(t,s)\}$. We omit the detailed definition here due to space limitations. Formulas are intended to represent sentences while terms are intended to represent arguments. The informal reading of the different kinds of formulas is the following: $B_a\varphi$ means "a believes that φ"; $D_{ab}\varphi$ means "b is an expert on topic φ according to a's opinion" or "b is the best φ-advisor of a" (taken from Huang, 1990); $t \gg \varphi$ means "argument t concludes φ (in a structural sense)" or "φ is a conclusion of argument t"; $Com^{\leq}(t,s)$ means "argument t is simpler than argument s".

With respect to the terms, the definition is taken from Baltag et al. (2012, 2014). The first clause states that there is one "atomic argument" c_φ for each formula φ of the language.[2] Atomic arguments can be informally understood as arguments that contain one sole premise identical to the conclusion, i.e., "It rains because it rains" or "It is Tuesday, therefore it is Tuesday". When they occur in more complex arguments, they can be seen as their premises. Both $+$ and $\cdot$ are the typical operations in justification logic,

---

[2]Baltag et al. (2012, 2014) called them *certificates*. This explains the original notation that we have decided to maintain.

but we reinterpret them in an argumentative reading. Therefore, a term of the form $[t+s]$ represents a complex argument composed by t and s without performing any logical inference. Besides, a term of the form $[t \cdot s]$ represents a complex argument resulting from combining the conclusions of t and s by the use of modus ponens.

This way of defining terms is somehow non-standard. Usually, in justification logic (see, e.g., Artemov & Fitting, 2016), terms are built departing from a set of constants and a set of variables. After that, primitive relations between constants and formulas are axiomatically assumed in what is called a *constant specification*. We have chosen to follow Baltag et al. (2012, 2014) because their way of defining terms matches better with our objective of modelling deductive arguments (understood as syntactic entities). There are two reasons to justify this advantage. First, the premises of a given argument t have the form $c_{\varphi_1}, c_{\varphi_2}, ..., c_{\varphi_n}$ and refer directly to concrete sentences $\varphi_1, ..., \varphi_n$ with a given structure, as it happens with informal arguments; and not to purely abstract entities, as Artemov's justification terms are. Second, it offers an elegant and systematic way to relate terms (arguments) to formulas (sentences) regarding the structure of the arguments.

Notation: For negated $\gg$-formulas, for instance $\neg c_p \gg q$, we add brackets and write $\neg(c_p \gg q)$ for improving readability.

2.2 Semantics

Before introducing the models in order to provide a meaning for the language, note that there are two types of formulas that intuitively bear a "syntactic meaning": $t \gg \varphi$ and $Com^{\leq}(t,s)$. The schema $t \gg \varphi$ asserts that argument t has the proper structural shape to support φ or, in other words, that φ is actually a conclusion of t. We follow Baltag et al. (2014) to capture the meaning of these formulas.

Definition 2 (Admissibility [Baltag et al., 2014]) *Define the meta-syntactic relation* $\gg \subseteq \mathcal{T} \times \mathcal{F}$. *More concretely*, $\gg$ *is the smallest relation satisfying the following clauses*[3] *(a)* $c_\varphi \gg \varphi$ *(b) If* $t \gg (\varphi \to \psi)$ *and* $s \gg \varphi$, *then* $[t \cdot s] \gg \psi$ *and (c) If* $t \gg \varphi$ *and* $s \gg \psi$, *then* $[t+s] \gg \varphi$. *Define also* conclusions$(t) := \{\varphi \in \mathcal{F} \mid t \gg \varphi\}$.

Two remarks should be made about the just introduced notion. First,

---

[3] We abuse of notation by using the same symbol, $\gg$, to denote both the operator of the object language and the syntactic relation in the metalanguage but this abuse will be harmless.

A Justification Logic for Argument Evaluation

clause (c) allows for arguments supporting multiple conclusions, for example $[c_p + c_q]$ is admissible for both p and q. Second, we have just included in our logic the two famous operators in justification logic, but other operators can be included without any technical problem as far as they preserve truth from the premises to the conclusion. For example, $t \triangle s$ interpreted as the result of combining t and s via introduction of conjunction or $\triangledown t$ as the result of applying introduction of disjunction to the conclusions of t.

As for measuring the complexity of arguments (i.e., interpreting $Com^\leq$-formulas), two simplicity criteria are taken into account: the quantity of atomic information (different atoms present in the premises) and the length of the arguments.

Definition 3 (Quantity of atomic information, length of an argument) *Define the function* atom $: \mathcal{T} \longrightarrow \wp(\Phi)$, *that maps each term t to a subset of Φ, concretely the subset of atoms that occur in its premises. Define also* length $: \mathcal{T} \longrightarrow \mathbb{N}$ *that maps each term to its length (the number of symbols it has excluding parenthesis). We omit here the detailed definition and just point out that* length$(B_a) = $ length$(D_{ab}) = 1$. *See, as examples,* atom$(c_{p \to (p \vee r \vee q)}) = \{p, q, r\}$ *and* length$(c_{B_a((c_{p \wedge q \to r} \cdot c_{p \wedge q}) \gg r)}) = 15$

Definition 4 (Model) *A model for $\mathcal{L}^{AE}$ is a tuple $M = \langle W, \{R_a\}_{a \in \mathcal{A}}, \mathcal{D}, [[\cdot]]\rangle$ where $W \neq \emptyset$ is a set of possible worlds; each $R_a \subseteq W \times W$ is an accessibility doxastic relation;* $[[\cdot]] : \Phi \longrightarrow \wp(W)$ *is a valuation map and* $\mathcal{D} : \mathcal{A} \times \mathcal{A} \times W \longrightarrow \wp(\mathcal{F})$ *is a best advisor map such that the following conditions hold:*

(1) $\varphi \in \mathcal{D}(a, b, w)$ iff $\neg \varphi \in \mathcal{D}(a, b, w)$; (neutrality)

(2) If $\varphi \in \mathcal{D}(a, b, w)$, then $\varphi \notin \mathcal{D}(a, c, w)$ for every $c \neq b$; (uniqueness)

(3) $\varphi \in \mathcal{D}(a, b, w)$ iff ($\varphi \in \mathcal{D}(a, b, w')$ for each w' such that wR_aw' and there is a $w' \in W$ such that wR_aw' (safety)

Condition (1) captures the intuitive idea that the best advisor of a regarding φ is also her best advisor regarding $\neg \varphi$. Condition (2) assures that the best advisor of a given agent is unique and it will avoid the possibility of having inconsistent preferences over arguments. Condition (3) establishes two necessary and sufficient requirements for a to have a best advisor with respect to φ. First, the notion of *best advisor* is understood as being doxastically determined, i.e., an agent has a best advisor if and only if she believes so and the second requirement holds. The second requirement is included for technical reasons, since if we allowed agents to have inconsistent beliefs together with beliefs about their advisors, condition (2) would be violated.

A *pointed model* is a pair $\langle M, w \rangle$ where M is a model and $w \in W$. Given a pointed model $\langle M, w \rangle$ we omit the brackets and denote it by M, w. Let us denote by $\mathcal{M}$ the class of all pointed models for $\mathcal{L}^{AE}$.

Definition 5 (Truth) *Let us define the* truth relation $\vDash$ *as the smallest relation* $\vDash \subseteq \mathcal{M} \times \mathcal{F}$ *that satisfies the following conditions (the missing clauses for atoms and logical connectives are the usual ones):*

$M, w \vDash Com^{\leq}(t, s)$ iff $\mathsf{length}(t) + |\mathsf{atom}(t)| \leq \mathsf{length}(s) + |\mathsf{atom}(s)|$
$M, w \vDash t \gg \varphi$ iff $\varphi \in \mathsf{conclusions}(t)$
$M, w \vDash B_a \varphi$ iff $\forall w' (w R_a w' \Rightarrow M, w' \vDash \varphi)$
$M, w \vDash D_{ab} \varphi$ iff $\varphi \in \mathcal{D}(a, b, w)$

Definitions of validity and local semantic consequence are standard (Blackburn, De Rijke, & Venema, 2002).

2.3 Proof system

Definition 6 (Proof system) *The theory for* deductive arguments evaluation, *abbreviated as* L^{AE}, *is defined in Figure 1.*

We propose this axiom system as a minimal theory, since no special properties are given to B_a with the exception of axiom *Safety about* D_{ab}. Nevertheless, extra restrictions (general consistency of beliefs, and different forms of introspection) may be added using standard modal techniques (Blackburn et al., 2002).

Theorem 1 (Soundness and strong completeness) *Let* $\varphi \in \mathcal{L}^{AE}$ *and* $\Gamma \subseteq \mathcal{L}^{AE}$; *we have* $\Gamma \vdash \varphi$ *iff* $\Gamma \vDash \varphi$.

Sketch of the proof. Soundness is straightforward by induction on the construction of deductions. As for completeness, it can be proven via canonical models. The definitions of deduction from premises, consistency and maximal consistency are standard (Blackburn et al., 2002). As for the operators $\gg$ and D_{ab} the reader can consult the work of Baltag et al. (2012) and Huang (1990), respectively. We just give the definition of the canonical model $M^c := \langle W^c, \{R_a^c\}_{a \in \mathcal{A}}, \mathcal{D}^c, [[\cdot]]^c \rangle$ such that:

1. $W^c := \{\Gamma \subseteq \mathcal{L}^{AE} \mid \Gamma \in \mathfrak{MC}\}$

2. $\Gamma R_a^c \Delta$ iff $\{\varphi \in \mathcal{L}^{AE} \mid B_a \varphi \in \Gamma\} \subseteq \Delta$

Axioms

All propositional tautologies

$\vdash t \gg \varphi$ whenever $t \gg \varphi$

$\vdash \neg(t \gg \varphi)$ whenever not $t \gg \varphi$ *(admissibility)*

$\vdash B_a(\varphi \to \psi) \to (B_a\varphi \to B_a\psi)$ *(normality of B_a)*

$\vdash D_{ab}\varphi \leftrightarrow D_{ab}\neg\varphi$ *(neutrality of D_{ab})*

$\vdash (B_a D_{ab}\varphi \wedge \neg B_a \bot) \leftrightarrow D_{ab}\varphi$ *(safety about D_{ab})*

$\vdash D_{ab}\varphi \to \neg D_{ac}\varphi$ for every $c \neq b$ *(uniqueness of the best advisor)*

$\vdash \mathsf{Com}^{\leq}(t, s)$ when $\mathsf{length}(t) + |\mathsf{atom}(t)| \leq \mathsf{length}(s) + |\mathsf{atom}(s)|$

$\vdash \neg\mathsf{Com}^{\leq}(t, s)$ when $\mathsf{length}(t) + |\mathsf{atom}(t)| \not\leq \mathsf{length}(s) + |\mathsf{atom}(s)|$
(complexity of information)

Rules

MP $\quad \vdash \varphi, \vdash \varphi \to \psi \implies \vdash \psi$

NEC $\quad \vdash \varphi \implies \vdash B_a\varphi$

Figure 1: Proof system

3. $[[p]]^c = \{\Gamma \in \mathfrak{MC} \mid p \in \Gamma\}$

4. $\mathcal{D}^c(a, b, \Gamma) := \{\varphi \in \mathcal{L}^{AE} \mid D_{ab}\varphi \in \Gamma\}$

where $\mathfrak{MC}$ is the set of all maximally consistent sets in L^{AE}. □

3 Searching the best argument for φ

Before sketching our answer to (*i*), let us contextualise further which kind of situations we aim to model. First, we suppose that the beliefs of the agent are the result of some previous epistemic processes, for instance, observations; therefore pure irrational beliefs are excluded. Second, we shall restrict our attention to a pure logical argument evaluation, so neither rhetoric features, like the convincing power of the arguments, nor dialectical ones, like the argumentative moves that are allowed in a moment of the dialogue, are taken into account.[4] Third, no time bounds are considered; agents will take as much time as they need –but always a finite amount– for their evaluations. Fourth, we remark that the goal of evaluation is fixed: justifying a given claim φ. In other words, we are looking for the best argument for φ (from the perspective of each agent). Finally, the resulting analysis has a clearly normative character, we do not talk about what epistemic agents actually do when they evaluate their arguments, but rather what they should do in order to have the best epistemic standards.

According to the analysis we are about to present, the notion of preference between deductive arguments is reducible to other notions. Concretely, it can be analysed attending to the following list: the *structural shape of the arguments*, the *acceptability of the premises* with respect to the agent's beliefs, the *expert's opinions* with respect to the premises, and lastly the *simplicity* of the involved arguments.

The intuitive picture behind our analysis is that, when assesing two arguments, an epistemic agent puts them to a test consisting of several filters or criteria. When some of these filters is passed by both arguments, we find situations in which the agent does not accept the state of evaluation and she demands more information. This will be expressed by an implication whose antecedent is the passing condition, and whose consequent is a question mark (?); symbol that will be used only for presentation purposes but not in the formal definition of preference. As the result of this process, the agent will weakly prefer t to s (prefer t as much as s) if t has passed at least as many filters as s. The remaining preference concepts (strict preference, indifference and incomparability) are defined as usual in the preference logic literature (Hansson, 2001; Hansson & Grüne-Yanoff, 2017). One way to do so is to take the notion of weak preference as primitive and define then the remaining notions. Assume $P_a^{\geq}(t, s, \varphi)$, standing for "agent a prefers at least as much argument t to argument s" (*weak preference*), we can define *strong preference*, $P^{>}$, using boolean connectives as follows $P_a^{>}(t, s, \varphi) := P_a^{\geq}(t, s, \varphi) \wedge \neg P_a^{\geq}(s, t, \varphi)$. We can also define a notion of

---

[4]See, e.g., (Groarke, 2017) for the classical distinction between logic, rhetoric and dialectics.

A Justification Logic for Argument Evaluation

indifference $P_a^{\approx}(t,s,\varphi) := P_a^{\geq}(s,t,\varphi) \wedge P_a^{\geq}(t,s,\varphi)$ and *incomparability* $P_a^!(t,s,\varphi) := \neg P_a^{\geq}(s,t,\varphi) \wedge \neg P_a^{\geq}(t,s,\varphi)$. In what follows, we focus on an analysis of weak preference and use the shorthands above when convenient.

The first filter of the assessment test is called *structural accuracy*. If a tries to justify φ using argument t, then it seems rather obvious that the first thing she should require of t is to have φ as its conclusion. Note that this criterion can be fulfilled by t or s independently or they can, also independently, fail on it. Therefore, when comparing t and s, agent a faces four cases. Let us see the principles we would like to capture in our definition of preference and later on we will argue for their intuitive appeal:

Case A.1 $(\neg(t \gg \varphi) \wedge \neg(s \gg \varphi)) \rightarrow P_a^{\approx}(t,s,\varphi)$

Case A.2 $(t \gg \varphi \wedge \neg(s \gg \varphi)) \rightarrow P_a^>(t,s,\varphi)$

Case A.3 $(\neg(t \gg \varphi) \wedge s \gg \varphi) \rightarrow P_a^>(s,t,\varphi)$

Case A.4 $(t \gg \varphi \wedge s \gg \varphi) \rightarrow$?

Case A.1 says that if none of the argument has φ as its conclusion, then the agent feels indifference about them since she considers them equally bad for justifying φ. Cases A.2 and A.3 say that when one argument has φ as its conclusion while the other does not, then the agent should strongly prefer the first argument to the second one. Finally, we see that A.4 presents the situation in which both argument syntactically conclude φ and so the agent demands more information to go on with her evaluation. We can then abbreviate $\rho_a^1(t,s,\varphi) := (t \gg \varphi \wedge s \gg \varphi)$ meaning "t and s satisfy the condition to pass the first filter".

In the second filter, we focus on the *acceptability* of the premises according to a's beliefs. It seems clear that a should prefer arguments whose premises are believed by her to arguments whose premises are not believed. With this idea in mind, we define a notion of acceptability following Baltag et al. (2012): $A_a t := \bigwedge_{c_\varphi \in sub(t)} B_a \varphi$ stands for "agent a accepts argument t", i.e., she believes each premise of the argument to be true. For our purpose, it will be useful to distinguish between doxastic non-acceptance ($\neg A_a t$) and doxastic rejection, so let us introduce $R_a t : \bigvee_{c_\varphi \in sub(t)} B_a \neg \varphi$, informally meaning "$a$ rejects argument t", i.e., she believes that at least one of the premises is false. Baltag et al. (2012, 2014) do not consider this notion since their intentions are different to ours. Note that, according to our semantics, it is not generally true that $\neg A_a t \rightarrow R_a t$; since one could be sceptic about some premise of t without believing that it is false. Now,

Alfredo Burrieza and Antonio Yuste-Ginel

let us also define the following shorthands (for reducing the number of relevant cases in this filter): $A_a^>(t,s) := (A_a t \land \neg A_a s) \lor (\neg R_a t \land R_a s)$:≈ "agent a considers t strictly better than s in doxastic terms" and $A_a^{\approx}(t,s) := \neg A_a^>(t,s) \land \neg A_a^>(s,t)$:≈ "a considers t and s equally good in doxastic terms". What are the relevant cases regarding this filter?

Case B.1 $(\rho_a^1(t,s,\varphi) \land A_a^>(t,s)) \to P_a^>(t,s,\varphi)$

Case B.2 $(\rho_a^1(t,s,\varphi) \land A_a^{\approx}(t,s)) \to$?

B.1 says that if the first filter has been passed and t is strictly more accepted than s, then strong preference for t is derived. B.2 says that if they are equally acceptable, the agent demands more information. Let us then abbreviate $\rho_a^2(t,s,\varphi) := (\rho_a^1(t,s,\varphi) \land A_a^{\approx}(t,s))$ meaning that both arguments satisfy the criterion to pass the second filter.

In the third filter the agent turns her attention to her *best advisor* or an *expert* on the topic φ. Once her beliefs have been found to be insufficient to decide which argument is better, she seeks help in other agents of the system. Notice that *the best advisor* is understood here in a broad sense (it could be a human being, a computer, an encyclopedia, etc.), i.e., the best advisor means the best source for the agent with respect to the topic. This idea is taken from Huang (1990), but some changes are made in order to fulfil our theoretical intentions.

The order of these two filters, the second and the third one, within the hierarchy of the test is not straightforward. We could think of situations in which it is epistemically better for an agent to first consult an expert and then she does some introspection to check her own beliefs (imagine, for example, cases where an agent had a medical issue but she is not a doctor). This would mean that the agent trusts more her advisor than herself and it looks perfectly reasonable. Nevertheless, we have chosen the present order due to a couple of reasons. First, since we have assumed that the beliefs of the agent are the result of some previous epistemic processes, it is also plausible to assume that she has certain confidence in them. Second, this order might imply less cognitive expense for the agent: she first consults her own database and then, if some information is missing, she looks for it in some external sources. This form of ordering the self confidence of the agents and the confidence they have in their advisors matches intuitively the belief revision strategy called *low-credibility advisor* by Huang and van Emde Boas (1991). Anyway, let us remind that this is just a tentative order that might be altered regarding the application and the context. Other levels of credibility

A Justification Logic for Argument Evaluation

could be included in our framework by simply changing the definition of preference that the reader will find at the end of this section. It is also possible to include different D_{ab}-operators in order to capture situations where the agents consider experts with different credibility.

Turning to the matter in hand, what are the possible cases in this phase of the assessment process? There are many combinations but they can be split into two main groups regarding preference. On the one hand (C.1), in the first group of situations we have that b is the best φ-advisor of a and b considers t strictly better than s (in terms of epistemic acceptance). The output in these cases will clearly be that a strictly prefers t to s: so strict preference is transferred from the advisor to the agent. On the other hand (C.2), there are situations where b considers t and s equally good or where a does not have any φ-advisor. The output here will be the demand for more information:

Case C.1 $(\rho_a^2(t,s,\varphi) \wedge D_{ab}\varphi \wedge A_b^{>}(t,s)) \to P_a^{>}(t,s,\varphi)$

Case C.2 $(\rho_a^2(t,s,\varphi) \wedge ((D_{ab}\varphi \wedge A_b^{\approx}(t,s)) \vee (\bigwedge_{b \in \mathcal{A}} \neg D_{ab}\varphi))) \to \ ?$

Unfortunately for our agent, it might be the case that her best advisor is not sufficiently informed to tell her what to do, since b (the advisor) could also consider t and s equally good, or that she has no advisor regarding φ, in symbols $\rho_a^3(t,s,\varphi) := (\rho_a^2(t,s,\varphi) \wedge ((D_{ab}\varphi \wedge A_b^{\approx}(t,s)) \vee (\bigwedge_{b \in \mathcal{A}} \neg D_{ab}\varphi)))$. However, we could still establish some criteria to make a decide.

Finally, in the fourth filter; the agent checks the *simplicity* of the arguments (and she should prefer, of course, the simplest).[5] It is not straightforward how to model the notion of simplicity in this context. Here, we take two different requirements into account: the way the information is structured in the arguments and the quantity of information they contain. It is not clear which of these two properties should be considered more important in the evaluation process, due to the fact that we could find intuitive counterexamples for the two orders. This is the main reason we have included both criteria in the truth clause for $Com^{\leq}$. The shorthands $Com^{\approx}(t,s)$ and $Com^{<}(t,s)$ are defined as expected. We can now see how this filter interacts with the preferences of the agent:

Case D.1 $(\rho_a^3(t,s,\varphi) \wedge Com^{<}(t,s)) \to P_a^{>}(t,s,\varphi)$

---

[5] See (Baker, 2016) for a general discussion about the uses and justification of simplicity criteria in philosophy and science.

Case D.2 $(\rho_a^3(t,s,\varphi) \wedge Com^{\approx}(t,s)) \to P_a^{\approx}(t,s,\varphi)$

Case D.1 says that, everything else being equal, if argument t is strictly simpler than s, then a should prefer t to s. Finally, if the fourth filter is passed by both arguments (i.e., they are equally complex), then agent a considers them equally good (case D.2). This filter concludes our analysis.

We are finally able to give a definition of the $P_a^{\geq}$ operator in the object language $\mathcal{L}^{AE}$. We would like to capture in this definition all the principles discussed in this section (cases A.1.-D.2.). The strategy for doing so is defining the preference operator as a disjunction where each disjunct is the antecedent of schemas A.1.-D.2. where strong preference appears as a consequent. Besides, a last disjunct is added to capture the situations where both arguments haven shown to be equally good with respect to all the criteria but one of them is simpler than the other.

Definition 7 (φ-preference between deductive arguments) *With the shorthands introduced in this section we define the preference operator $P_a^{\geq}$ as follows:*

$$\begin{aligned}P_a^{\geq}(t,s,\varphi) := & (\neg(t \gg \varphi) \wedge \neg(s \gg \varphi)) \vee (t \gg \varphi \wedge \neg(s \gg \varphi)) \vee \\ & \vee (\rho_a^1(t,s,\varphi) \wedge A_a^{>}(t,s)) \vee \\ & \vee (\rho_a^2(t,s,\varphi) \wedge D_{ab}\varphi \wedge A_b^{>}(t,s)) \vee \\ & \vee (\rho_a^3(t,s,\varphi) \wedge Com^{\leq}(t,s)) \end{aligned} \quad (P)$$

It is interesting to note that the preference operator has three of the most often discussed properties in the logical literature about preferences (Hansson & Grüne-Yanoff, 2017). The proof of the next proposition is simple but long, so we state it without proof because of space limitations.

Proposition 1 (Properties of preference) *Let $a \in \mathcal{A}$; $\varphi \in \mathcal{F}$ and $t, s, r \in \mathcal{T}$, the preference operator induces a total preorder on $\mathcal{T}$, i.e., the following schemas are valid:*

- $\models P_a^{\geq}(t,t,\varphi)$ (reflexivity)
- $\models P_a^{\geq}(t,s,\varphi) \wedge P_a^{\geq}(s,r,\varphi) \to P_a^{\geq}(t,r,\varphi)$ (transitivity)
- $\models P_a^{\geq}(t,s,\varphi) \vee P_a^{\geq}(s,t,\varphi)$ (connectedness)

4 Closing words and future work

Summing up, this paper has introduced an innovative logical system, called the logic AE, that combines several existing tools from three main sources: doxastic logic, logics for belief dependence and justification logic. The aim of creating such a system has been modelling a range of scenarios where different agents asses a couple of deductive arguments in order to decide which one is better to support a given claim. From a general perspective, it represents a small step towards studying the interactions between justification logic and argumentation theory.

There are two key ideas behind our analysis. First, the result of argument evaluation can be expressed in the new logic as a preference relation between arguments. Second, this preference relation can be reduced to a list of other argumentative and epistemic notions. Concretely, we have reduced the preference between deductive arguments to the following hierarchic list: the structural shape of arguments (do they actually conclude the claim φ?); the doxastic acceptance of the premises (does the agent believe that the premises are true?); the view of an expert or distinguished advisor (what does the expert think of the acceptability of the arguments?) and finally the simplicity of the arguments (which argument is simpler?).

We would like to mention some further work that has been already developed but has been left out due to space limitations. First, the decidability of a slight modification of AE and some extensions of it (assuming extra properties of B_a, D_{ab} and including both a $S4$ and a $S5$ knowledge operator) can be proven via the strong finite model property. Second, we have identified an interesting face of the famous logical omniscience problem regarding the defined preference operator: if an agent is trying to justify a claim φ, then the atomic argument c_φ is always at least as good as any other. In other words, the schema $P_a^\geq(c_\varphi, t, \varphi)$ is a valid formula in the logic AE. This fact can be found counterintuitive and undesirable in many situations. The problem can be solved by the introduction of a term awareness operator (following Baltag et al., 2012).

As for the future work, we just mention two promising lines. First, it seems interesting to extend the present system using dynamic epistemic logic tools in order to study how different informational updates interact with argument evaluation. Second, it would be interesting to explore ways to get rid of our main contextual assumption, i.e., restricting the analysis to deductive arguments, by including other types of arguments.

References

Artemov, S. (2008). The logic of justification. *The Review of Symbolic Logic*, *1*(4), 477–513.

Artemov, S., & Fitting, M. (2016). Justification logic. In E. N. Zalta (Ed.), *The Stanford Encyclopedia of Philosophy* (Winter 2016 ed.). Metaphysics Research Lab, Stanford University. https://plato.stanford.edu/archives/win2016/entries/logic-justification/.

Artemov, S., & Nogina, E. (2005). Introducing justification into epistemic logic. *Journal of Logic and Computation*, *15*(6), 1059–1073.

Baker, A. (2016). Simplicity. In E. N. Zalta (Ed.), *The Stanford Encyclopedia of Philosophy* (Winter 2016 ed.). Metaphysics Research Lab, Stanford University. https://plato.stanford.edu/archives/win2016/entries/simplicity/.

Baltag, A., Renne, B., & Smets, S. (2012). The logic of justified belief change, soft evidence and defeasible knowledge. *Logic, Language, Information and Computation*, 168–190.

Baltag, A., Renne, B., & Smets, S. (2014). The logic of justified belief, explicit knowledge, and conclusive evidence. *Annals of Pure and Applied Logic*, *165*(1), 49–81.

Blackburn, P., De Rijke, M., & Venema, Y. (2002). *Modal Logic: Graph. Darst* (Vol. 53). Cambridge University Press.

Groarke, L. (2017). Informal logic. In E. N. Zalta (Ed.), *The Stanford Encyclopedia of Philosophy* (Spring 2017 ed.). Metaphysics Research Lab, Stanford University. https://plato.stanford.edu/archives/spr2017/entries/logic-informal/.

Grossi, D., & van der Hoek, W. (2014). Justified beliefs by justified arguments. In *Proceedings of the Fourteenth International Conference on Principles of Knowledge Representation and Reasoning* (pp. 131–140).

Hansson, S. O. (2001). Preference logic. In D. M. Gabbay & F. Guenthner (Eds.), *Handbook of philosophical logic* (pp. 319–393). Springer.

Hansson, S. O., & Grüne-Yanoff, T. (2017). Preferences. In E. N. Zalta (Ed.), *The Stanford Encyclopedia of Philosophy* (Winter 2017 ed.). Metaphysics Research Lab, Stanford University. https://plato.stanford.edu/archives/win2017/entries/preferences/.

Huang, Z. (1990). Logics for belief dependence. In *International Workshop on Computer Science Logic* (pp. 274–288).

Huang, Z., & van Emde Boas, P. (1991). *Belief dependence, revision and persistence*. Instituut voor Taal, Logica en Informatie.

Renne, B. (2012). Multi-agent justification logic: Communication and evidence elimination. *Synthese, 185*, 43–82.

Shi, C., Smets, S., & Velázquez-Quesada, F. (2017). Argument-based belief in topological structures. *Electronic Proceedings in Theoretical Computer Science, 251*.

Shi, C., Smets, S., & Velázquez-Quesada, F. R. (2018). Beliefs supported by binary arguments. *Journal of Applied Non-Classical Logics*, 1–24.

van Benthem, J., Fernández-Duque, D., & Pacuit, E. (2012). Evidence logic: A new look at neighborhood structures. *Advances in modal logic, 9*, 97–118.

van Benthem, J., Fernández-Duque, D., & Pacuit, E. (2014). Evidence and plausibility in neighborhood structures. *Annals of Pure and Applied Logic, 165*(1), 106–133.

van Benthem, J., & Pacuit, E. (2011). Dynamic logics of evidence-based beliefs. *Studia Logica, 99*(1), 61.

Alfredo Burrieza
University of Málaga, Department of Philosophy
Spain
E-mail: burrieza@uma.es

Antonio Yuste-Ginel
University of Málaga, Department of Philosophy
Spain
E-mail: antonioyusteginel@gmail.com

On the Nature of Combinatorial Proofs

SERENA DELLI

Abstract: In this paper I am going to analyse the graph-theoretical formalism of combinatorial proofs, and to discuss the syntactic or semantic nature of this calculus.

Keywords: proof theory, combinatorial proofs, syntax-free formalisms

1 Introduction

During the mid-1990 R. Thiele, a historian of science, discovered from an old archive in Göttingen some notes concerning a formulation of a 24th problem to be added to Hilbert's famous talk in Paris (Thiele, 2005; Thiele & Wos, 2002). The problem asks for the simplest proof of any theorem. Hilbert's 24th problem opened new lines of research in philosophy and foundation of mathematics, as well as in proof theory. One of the main questions concerns the *identity of proofs* (Straßburger, 2007). Proof theory, unlike other mathematical disciplines, seeks a clear and well-defined notion of identity for formal proofs. This is due to what Girard called *logical bureaucracy*, which allows the formalisation of the same argument in different ways. For this reason Straßburger writes that "proof theory is not a theory of proofs but a theory of proof formalisms" (Straßburger, 2017). The following two proofs provide and example on how trivial rule permutations change the structure of derivations, while the argument is preserved.

$$
\cfrac{\cfrac{\cfrac{\overline{\vdash p, \bar{p}} \text{ ax.}}{\vdash p, \bar{p}, q} \text{ weak.}}{\cfrac{\vdash p \land p, \bar{p}, \bar{p}, q}{\vdash p \land p, \bar{p}, q} \text{ cont.}} \overline{\vdash p, \bar{p}} \text{ ax.}}{\vdash p \land p, \bar{p}, q} \land \qquad \cfrac{\cfrac{\cfrac{\overline{\vdash p, \bar{p}} \text{ ax.} \quad \overline{\vdash p, \bar{p}} \text{ ax.}}{\vdash p \land p, \bar{p}, \bar{p}}}{\cfrac{\vdash p \land p, \bar{p}}{\vdash p \land p, \bar{p}, q} \text{ weak.}} \land}{}
$$

This makes it clear that a different approach to proof theory was necessary. New, lighter, bureaucracy-free formalisms have been designed. Girard's proof-nets constitute one of the first attempts in this direction (Girard,

1987), but his formalism only well captures a small fragment of classical logic. A considerable step forward has been made by Dominic Hughes, who introduced the notion of combinatorial proofs. This approach gets rid of the logical language, i.e., the syntax, replaces formulae with their associated graphs, and generates a syntax-free calculus. Validity for combinatorial proofs is characterised by expressing proofs in terms of graphs and homomorphisms. Intuitively, a proof, in the sense of combinatorial proofs, is a projection of an axiomatic structure into the provable formula. The purpose of this paper is to defend the epistemic status of these proofs. In the next section I am going to give you the technical details, which will be followed by a more philosophical argument. This argument makes use of two main concepts: the dichotomy of syntax and semantics and the epistemic status of diagrammatical reasoning. In the conclusion I will defend the claim that combinatorial proofs are a syntactic formalism that deserve to be called a proof system.

2 Combinatorial proof

Combinatorial proofs, as they were presented by Hughes (2006a, 2006b), are a graph-theoretical representation of classical logic, where its syntax is hidden. This formalism is based on three main notions: the graph of a formula, a particular notion of homomorphism (called skew fibration), and the notion of nicely coloured graph. In the next paragraphs I will introduce all these concepts, and I will give a glimpse on how this formalism can help us with the problem of identity. Here, the formulae are composed by atomic occurrences, and possibly the negations, and the primitive connectives are just conjunction ($\wedge$), disjunction ($\vee$). Since we are working in classical logic, we can rewrite implication as a combination of negation and disjunction namely, $A \to B \iff \neg A \vee B$.

2.1 The graph of a formula

We will now see how to associate any formula to its cograph. The aim is to find a way to represent conjunction and disjunction by features in a graph. Intuitively, conjunctions create a tighter bound between formulae than disjunctions. This is rendered by drawing a link just between all the terms of conjunctive formulae.

Definition 1 *Given a formula A the associated graph of the formula G_A*

On the Nature of Combinatorial Proofs

is defined as follow: the vertices are simply all the atomic occurrences[1] *in A; the edges are defined inductively: if A is the form* $B \wedge C$, *then* G_A *is obtained by drawing an edge between the vertices of* G_B *and* G_C *and no other edge appears in* G_A

In particular, the graph of a formula is a *cograph*[2].

Notice that we will use $p, q, r, \ldots$ to denote atomic occurrences, and $\bar{p}, \bar{q}, \bar{r}, \ldots$ to denote their negations.

For example, consider the a fortiori law ($\cong \neg p \vee \neg q \vee p$) and $[p \to (q \to r)] \to [(p \to q) \to (p \to r)]$ ($\cong (p \wedge q \wedge \neg r) \vee (p \wedge \neg q) \vee (\neg p \vee r)$). The associated graph are respectively:

$$\bar{p} \quad \bar{q} \quad p$$

$$q$$
$$\diagup\diagdown$$
$$p \text{——} \bar{r} \quad r \quad \bar{p} \quad p \text{——} \bar{q}$$

It is possible to prove that given a formula A there is a unique graph G_A associated to it and given a cograph G = (V, R), there is a formula A_G associated to it. The formula A_G is unique, up to commutativity and associativity of $\wedge$ and $\vee$. For the proofs and more technical details see (Carbone, 2010).

2.2 Nicely coloured graph

A very peculiar notion of combinatorial proofs is the *nicely coloured graph*.

Definition 2 *A graph (V, E) is coloured if V carries an equivalence relation* $\sim$ *such that* $v \sim w$ *only if* $(v, w) \in E$. *Each equivalence class is a colour class.*

We are going to see first what is coloured graph and the what is a nicely coloured graph.

Definition 3 *A colouring of a graph (V,E) is nice if every coloured class has at most two vertices and no union of two-vertex coloured classes* $[v_1, v_2]$ *and* $[w_1, w_2]$ *induce a matching. A matching is induced if there exists exactly the edges* $(v_1, w_i) \in E$ *and* $(v_2, w_j) \in E$ *for i=1 and j=2, or i=2 and j=1.*

Consider the following graphs[3]:

---

[1] Here with atomic occurrences we mean occurrences of an atomic formula, i.e., propositional variable.

[2] A graph is a cograph if it can be constructed from isolated vertices by disjoint union and join operations.

[3] In this and in the following examples, different shapes mean different colours as well as same shape indicates that the two nodes belong to the same coloured class.

The first graph on the left is not nicely coloured since it induces a matching, the middle one does not have the two-vertices property. Only the last one is a nicely coloured graph.

2.3 Skew fibration

Intuitively, we can see combinatorial proof as a projection of an axiomatic structure (the nicely coloured graph) into the formula we want to prove (the graph of the formula). In order to get the formal definition of combinatorial proof, we still miss a notion able to characterise this sort of projection. The morphism we are looking for is called *skew fibration*:

Definition 4

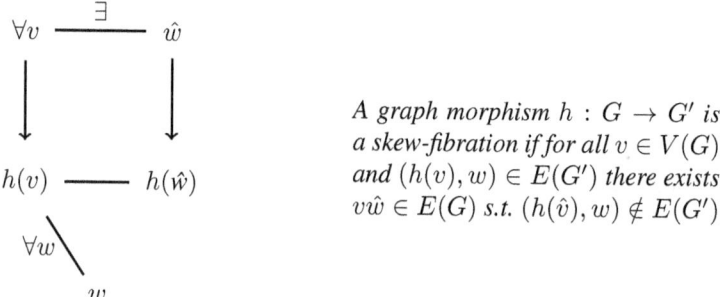

A graph morphism $h : G \to G'$ is a skew-fibration if for all $v \in V(G)$ and $(h(v), w) \in E(G')$ there exists $v\hat{w} \in E(G)$ s.t. $(h(\hat{v}), w) \notin E(G')$

Skew fibration is a lax form of graph fibration, and it just a variation on the usual definition of fibration in topology and category theory. With this final notion we can give the definition of combinatorial proof:

Definition 5 *A combinatorial proof of a formula A is a skew fibration $h : C \to G_A$ from a nicely coloured cograph C to the graph G_A such that, every coloured class is axiomatic, namely given $(v, w) \in V_C$ which belong to the same coloured class $h(v)$ and $h(w)$ are labeled with dual literals.*

Considering $[p \to (q \to r)] \to [(p \to q) \to (p \to r)]$ we have the following combinatorial proof:

On the Nature of Combinatorial Proofs

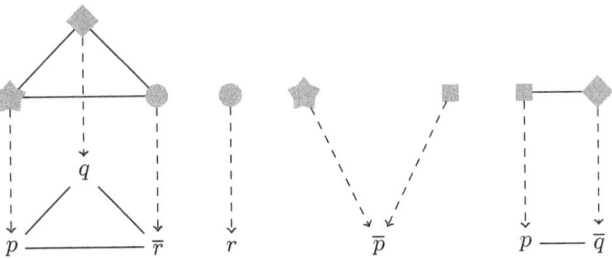

2.4 Combinatorial proofs and the problem of identity

Recall the starting point was the problem of identity, i.e., the two proofs in the introduction failed to be recognised as the same proof. We introduced combinatorial proofs as a way to uncover the structure of a derivation so far hidden by syntax, in order to identify all the proofs that use the same argumentation, i.e., have the same structure. What is left now is just to define a procedure to associate combinatorial proofs to the derivations in sequent calculus. But this is an easy task. Indeed we can associate a combinatorial proof at each step of the derivation in the following way:

Definition 6 *We can consider, without loss of generality, a sequent calculus á la Tait for classical logic (for details see Tait, 1968). The map is defined inductively as follows: (1) Each axiom is associated a two-vertex coloured class. Since axioms are of the form ⊢ p, ¬p, we have the axiomatic property; (2) when there is a conjunction rule, we draw edges between the vertices of the coloured graphs associated with the terms of the conjunction; (3) disjunction do not modify the coloured graph, neither the homomorphism; (4) weakening does not affect the coloured graph, just the graph of the formula; (5) when there is a contraction, all the coloured vertices of the contracted formulae are mapped into one, single atomic formula;*

Thus, we can claim that two proofs are the same when they have the same combinatorial proof. Drawing the associated combinatorial proofs on the derivations presented in the introduction we obtain:

$$\cfrac{\cfrac{\vdash p,\bar{p} \;\;\overset{\bullet\;\bullet}{}}{\vdash p,\bar{p},q} \text{ weak.} \qquad \vdash p,\bar{p} \overset{\circ\;\circ}{} }{\cfrac{\vdash p \wedge p, \bar{p},\bar{p},q}{\vdash p\wedge p,\bar{p},q} \text{ con.}} \wedge \qquad \cfrac{\cfrac{\vdash p,\bar{p}\;\;\overset{\bullet\;\bullet}{} \qquad \vdash p,\bar{p}\;\;\overset{\circ\;\circ}{}}{\vdash p\wedge p,\bar{p},\bar{p},} \wedge }{\cfrac{\vdash p\wedge p,\bar{p}}{\vdash p\wedge p,\bar{p},q} \text{ weak.}} \text{ con.}$$

As you can see, the combinatorial proofs are exactly the same, i.e., these two proofs differ only by trivial permutation rules. But this also means, they use the same argument and thus are the same proof.

3 Are combinatorial proofs "real" proofs?

The first step in our discussion should be the attempt to answer a fundamental question: *what is a proof?* This is a reasonable question, although very complicated to answer. It is easy to claim that proofs belong to syntax. Thus, we should start analysing what is syntax and what is semantics. The plan of this section is to underline some of the essential features of syntax and semantics and then study the notion of *formal proof*. Keep in mind that this is preparatory to the next section, where we will apply our arguments to combinatorial proofs.

3.1 Syntax vs semantics

Logics are usually described by their syntax and their semantics, but how can we define these concepts? Very naively we can state that syntax defines a language and tools to derive formulae in the given language; while semantics gives structures which determine the validity of the formulae. Syntax and semantics generate tools to identify, given a certain language, two special classes of formulae: the former the class of provable formulae, while the latter the one of valid formulae. The objects studied by syntax are proofs, while for semantics the objects are, mostly, models. The aim of this section is to point out some differences between these two subjects.

Given a logic, what we want is to distinguish valid from invalid formulae. Semantically, we say that a formula is valid in a model, if *there are no counterexamples*. Semantics typically speaks about the truth of formulae and identifies the class of true formulae. While on the syntactic side,

On the Nature of Combinatorial Proofs

to certify the validity of a formula we need *a proof*. What syntax picks out is the class of provable formulae. If we think in terms of invalid argument, semantically the invalidity of a formula is given by the *existence* of a counterexample and syntactically by the *non-existence* of a proof.

Due to universal quantifier involved in the definition of semantic validity, there is an infinitary nature to semantics. Affirming that a certain formula is a tautology ($\models A$) means that A is true in *all* models. But how can we verify all the possible models? Semantically speaking, there are some algorithmic ways to decide if a certain formula is a tautology, i.e., the method of truth tables. But there are two main problems: the complexity grows really quickly and it works only for propositional logic. That is why having a syntax and a proof system is *crucial*.

Syntax gives rules to determine what follows from what. It gives an algorithmic way to derive, from certain axioms, more complex formulae. Proofs witness the validity of formulae, and beside the possibility of having an infinite number of proofs for the same formula, only one derivation is needed. This existential quantifier makes syntactic validity very handy. A direct consequence of this is the complexity of checkability: on the semantic side, we usually have complexity in exponential time, while with syntactic formalisms we request polynomial time.

But now the question is: are these two notions of validity extensionally the same? Clearly, this is a potential problem: we have two different definitions for the same notion which might generate two different classes of formulae. Recall the theorems of soundness and completeness. The former says $if\ \Gamma \vdash \phi\ then\ \Gamma \models \phi$, namely if you can prove ϕ then ϕ is true. Or, from a different perspective, you are not able to derive any false formula. The latter $if\ \Gamma \models \phi\ then\ \Gamma \vdash \phi$ namely, if we start with valid premises, the inference rules do not allow an invalid conclusion to be drawn. Completeness guarantees all valid formula can be derived from the axioms and the inference rules. So there are no valid formulae that we can not prove.

I believe that these two theorems witness the blurriness of the boundary between syntax and semantics. They not just connect these apparently dichotomous notions, but they also testify that the classes described by syntax and semantics are extensionally the same. Other examples of this blurred line are infinitary proofs in sequent calculus, or tableaux formalism. About the latter, there is an ongoing debate on whether or not it should be consider a syntactic formalism of a sort of semantics. Even on proof nets there have been concerns on the nature of the formalism, generating an analogous discussion.

To better understand the two different notions modelled by syntax and semantics, we can consider an example: try to derive from $\neg(p \vee q)$, $\neg p$. Syntactically, we start our derivation having $\neg(p \vee q)$. In order to derive $\neg p$ in a given propositional calculus, we will have *to follow a set of transition steps*. The crucial one will be the application of *De Morgan laws*, which allow the following rewrite: $\neg p \wedge \neg q$. Then, applying some conjunction elimination rule you can obtain $\neg p$. Semantically, we can affirm that $\neg p$ is a logical consequence of because *all the ways of assigning values* to p and q that make $\neg(p \vee q)$ true, also make $\neg p$ true.

Using truth table is easy to verify that whenever $\neg(p \vee q)$ is true (1), even $\neg p$ is true. This means that $\neg p$ is logical consequence of $\neg(p \vee q)$; in symbols: $\neg(p \vee q) \models \neg p$.

As we said above, syntactic and semantic entailment are extensionally the same for propositional and first-order logic; then we can use the tools from syntax, which are much easier to handle, to decide whether or not a formula is a tautology. Indeed if proofs have certain properties, i.e., in the sequent calculus the end sequent has an empty antecedent, then we can affirm that the end formula is a tautology. Nevertheless, a proof system, in order to be useful, needs to have a good criterion of checkability, namely given a proof, there is a way to check if it is a correct proof of the formula.

> "[S]emantic methods tend to give one a better understanding" but they go on, "[they are] based on universal quantification over that mysterious totality, the class of all models (there are infinitely many models, and models themselves can be infinitely large). The notion of meaning that we use in the syntactic approach is more instrumental: the meaning of some part of the sentence lies in the conclusions which, because precisely that part appears at precisely that place, can be drawn from the sentence [...]". (Gamut, 1991, quoted in Szabolcsi, 2007)

This quote well summarises the problem of universal quantification in semantics, while introduces a new point: *the problem of meanings*.

I believe both syntax and semantics express a notion of meaning, but in a very different way. Syntax gives the meaning of the use, and it is completely determined by the rules of the calculus. It is possible to argue connectives are both softly and hardly defined. Softly in the sense that I can always give a different definition for all connectives; hardly because once the meaning is fixed by the rules, it can not change, and there is no space left for personal interpretations. Already Gentzen suggested that the meaning of connectives

On the Nature of Combinatorial Proofs

lies in the rules of the calculus:

> The introductions represent, as it were, the "definition" of the symbols concerned, and the eliminations are no more, in the final analysis, than the consequences of these definitions. This fact may be expressed as follows: In eliminating a symbol, we may use the formula with whose terminal symbol we are dealing only "in the sense afforded by the introduction of that symbol" (Gentzen, 1970, pp. 80–81)

But at the same time we have a semantics that gives a general meaning to the connectives. For example we can analyse the conjunction case. Semantically, $A \wedge B$ is true iff A is true *and* B is true. Syntactically, according to Gentzen's quote, the meaning of $\wedge$ can be given by the introduction rule.

I strongly agree that syntactic rules give some sort of meaning to connectives, but at the same time I have some hesitations on believing that is *completely* determined by the introduction rules.

What if we consider sequent calculus instead of natural deduction? Does the conjunction in sequent calculus mean something different? As it is well known, it is possible to formulate several sequent calculi for the same logic. The main issue is to prove they are equivalent. But then a question arises: do different formalisations of a chosen proof system give different meanings to connectives? For example, we can consider the two following rules for implication:

$$\frac{\Gamma \Rightarrow \Delta, A \quad B, \Gamma \Rightarrow \Delta}{A \to B, \Gamma \Rightarrow \Delta} \to L \qquad \frac{\Gamma \Rightarrow \Delta, A \quad B, \Pi \Rightarrow \Sigma}{A \to B, \Gamma, \Pi \Rightarrow \Delta, \Sigma} \to L$$

We can easily derive one from the other in the following way[4]:

$$\frac{\dfrac{\Gamma \Rightarrow \Delta, A}{\Gamma, \Pi \Rightarrow \Delta, \Sigma, A}\,\text{WR; WL} \quad \dfrac{B, \Pi \Rightarrow \Sigma}{\Gamma, \Pi \Rightarrow \Delta, \Sigma, B}\,\text{WR; WL}}{A \to B, \Gamma, \Pi \Rightarrow \Delta, \Sigma} \to L$$

Since these two rules can be derive one by the other, we could say that they express the same meaning. Nevertheless, if we look at the two derivations, we can notice that we make an essential use of *structural rules*. Then we could wonder if is it just the operational rules that give meaning to a connective, or do the structural rules influence it? What if we take operational

---
[4]The vice versa is obtained in a similar way.

rules with some build-in structural rules? All of these questions are generated by the syntactical bureaucracy, because of the wild diversification that syntax creates. I believe that semantics gives a core meaning, some essential and intuitive notion of the connective; while syntax defines the use and the meaning in that context. It is not possible to have a complete reduction of the meaning in proof theoretic terms.

The essential meaning, the primitive meaning, will always be given by semantics. Syntax enriches it, gives more properties and details, but it also allows diversifications which make problematic to recognise when two proofs are the same. I think semantics works as a minimum of meaning, in the sense that gives us just the necessary and sufficient properties of a connective. It allows us to recognise among the various proof systems and formalisations if the connectives involved mean the same. In the last paragraph we will see which role combinatorial proof plays in this context. There is an even more strong position, which believes in a stable meaning of the connectives across different logics.

3.2 What is a proof and diagrammatical reasoning

It still remains to answer the main question: *What is the epistemic value of Combinatorial Proofs?* Or, in different words, are Combinatorial Proofs real "proofs"? For this purpose, I want to discuss the notion of proof and to introduce the notion of visual thinking, since is clear the "not canonical" structure of combinatorial proofs. In fact, they are just graphs and morphisms; making them closer to combinatorics than to proof theory. What I am going to argue in this section is that the intuitive concept of proof does not describe all the possible forms of formal proof.

The role of diagrams and their epistemic value plays a central role in the philosophy of mathematical practice, and one of its main questions is if diagrammatical proofs are just tools of understanding, or if they are *proofs*. At the beginning of the last century, the prominent idea was to consider the diagrams as just a vehicle to help to understand a concept or to find a proof or a relation, but they had no epistemic value. For example, Hilbert in lectures on the foundations of geometry writes:

> A system of points, lines, planes is called a diagram or figure. The proof can indeed be given by calling on a suitable figure, but this appeal is not at all necessary. ... A theorem is only proved when the proof is completely independent of the dia-

On the Nature of Combinatorial Proofs

gram. *The proof must call step by step on the preceding axioms.* (Hilbert, 1894, p. 11, quote from Mancosu, 2005).

Here is evident how diagrams can be used to discover concepts, relations or even proofs, but they have no justification power. Proofs, *real*, epistemically valuable proofs need to be led by a linguistic characterisation of proofs as a sequent of sentences:

> We will acknowledge only those proofs in which one can appeal *step by step to preceding propositions and definitions*. If for the grasp of a proof the corresponding figure is indispensable then the proof does not satisfy the requirements that we imposed on it. These requirements are fulfillable; in any complete proof the figure is dispensable [...]. (Pasch, 1882/1926, p. 90, quote from Mancosu, 2005).

I want to underline two aspects that emerge in the previous quotes: (1) first of all, a proof is a step-by-step procedure; (2) proofs have a strong linguistic component. These are our two-first criteria on which we will base an intuitive notion of proof.

To better characterise this concept, I want to analyse another two quotes:

> A proof is a sequence of formulae each of which is either an axiom or follows from earlier formulae by a rule of inference. (Hilbert, 1930)

> Prior to the invention of formal logic, a proof was any convincing argument. [...] mathematicians write rigorous proofs, i.e. proofs in whose soundness the mathematical community has confidence. (Bundy, Jamnik, & Fugard, 2005)

From the former quotes, we can enrich our definition of proof. A proof is not only a *step-by-step* procedure, but at any step I write an axiom or I derive something new from the previous steps by applying a rule. I always know what I am doing and which axioms, steps or rules are involved. Moreover, a proof needs to be formal and objective. They are formulated in a way that any other person (with sufficient background) is able to understand and agree. Furthermore, proofs have an explanatory power. Reading a proof we get convinced on the correctness of the reasoning. I think these are all genuine properties of what intuitively we call a proof. Then we can add some more points to our naive notion of proof: (3) At each step formulae are axioms or obtained by previous step by applying rules; (4) proofs are rigorous or formal and (5) they have an explanatory power.

Then what is the role of diagrams? Recalling the two quotes from Hilbert and Pasch, we can add a "bonus" point on the nature of diagrams and their relation with proofs: diagrams or figures give just an intuition of the theorem. Using Giaquinto words,

> Thinking in terms of figures was valued as a means of facilitating grasp of formulae and linguistic text, but only reasoning expressed by means of formulae and text could bear any epistemological weight. (Giaquinto, 2008)

Historically speaking, the return to visual thinking is strongly influenced by the development of computer science and its interaction with mathematics. The advancement of computer graphics has allowed researchers to display mathematical notions in the forms of graphs. This is true even for logicians. With the development of theoretical computer science, the habits of representing formulae as a graph is become quite normal. Examples of this approach in logic are all those complexity results obtained by using graph representations of formulae combined with graph's properties. Combinatorial proofs are another result of this view of formulae in graph theoretical terms, and the interaction of logic with theoretical computer science. Nevertheless, mathematicians still considered visual representations as merely heuristic or as a support to find or verify new proofs:

> Many modern fields of mathematics admit visual presentations which do not, of course, claim to be logically rigorous but, on the other hand, offer a prompt introduction into the subject matter. (Fomenko, 2012)

It is clear that diagrammatical reasoning does not fit our naive idea of proof. One might think that proofs are just that, but I think the intuitive concept of proofs is not enough to characterise such a complex notion. We should at least divide proofs in formal proofs and proofs in terms of social interactions. Of course, here we are interested in formal proofs.

When we talk about formal proofs, we refer to mostly syntactic objects inscribed into a formal theory, generally a mathematical theory. But even on the formal side, you can have a huge variety of proofs. In logic, proofs are constructed by a very syntactical object, the formal derivation, and by the theory they belong to. Borrowing again Gianquinto's words:

On the Nature of Combinatorial Proofs

What about Tennant's claim[5] that a proof is 'a syntactic object consisting only of sentences' as opposed to diagrams? A proof is never a syntactic object. A formal derivation on its own is a syntactic object but not a proof. ... A formal derivation plus an interpretation and soundness proof can be a proof of the derived conclusion. But one and the same soundness proof can be given in syntactically different ways, so the whole proof, i.e. derivation + interpretation + soundness proof, is not a syntactic object. Moreover, the part of the proof which really is a syntactic object, the formal derivation, need not consist solely of sentences; it can consist of diagrams, as Miller's example shows. (Giaquinto, 2008)

The final question is if is possible to define some formal system for diagrammatical reasoning, and the answer is more than positive. Indeed, there are examples of formal diagrammatic proofs. Miller system for euclidian geometry is one; but there are plenty of other examples as the system of Venn diagrams developed by Ship or the diagrammatic system of geometry studied by Luengo, and Peirce system for classical logic.

What matters to me is to point out how the syntactic proof can be any well defined syntactic object given in a certain sound mathematical theory. This can be the Miller's formulation of geometry, proof nets, Peirce diagrams or combinatorial proofs. These are all syntactic objects that have all the rights to be part of the vast family of formal proofs. What gives epistemic value to the syntactic objects is not the form, but the theory. Some of them have a handy readability properties, like sequent calculus or natural deduction, where the line of reasoning is always very clear and there are not obscure passages. But some of them do not. An example is Boole calculus, where there are some steps that are completely uninterpretable. Of course we can always argue on what personally are the essential features of proofs, for some of us will be the explanatory power, for others the step-by-step structure, but I believe there are no technical reasons to not admit all of these very different derivations into the study of proofs.

---

[5]This is the claim he is citing: "[The diagram] has no proper place in the proof as such. For the proof is a syntactic object consisting only of sentences arranged in a finite and inspectable array."

4 Conclusion

What I tried to do in this paper is to present some thoughts that me and my research group discussed.[6] What is left to do is to bring back this discussion to our starting point: the combinatorial proofs. I think is already clear what is my opinion, and the answer to the title. But I also think is useful to analyse the arguments we debated above in terms of this unusual formalism. And this the aim of this last section.

If we were just looking for a technical reason to declare this formalism as semantic or syntactic, we could end this discussion in one paragraph: we have a theorem which guarantees that the checkability of combinatorial proofs is polynomial time (Cook & Reckhow, 1979; Hughes, 2006b). This crucial result would allow us to characterise combinatorial proofs as a syntactic formalism with no needs of further discussion. Notice that having a result like this is something sensational, and this is what makes this formalism such an amazing tool. Indeed, one of the main problems with all the other bureaucracy-free formalisms was the impossibility to define a checkability-principle in polynomial time. This made those formalisms very difficult to use, because on one side we were able to generate a syntactic object (a proof), but on the other side we were not able to assure its correctness.

Then, if we were looking just for a purely formal way of defining syntax and semantic, we could have our answer. But, unfortunately, if we look at the problem from other sides, the question becomes trickier.

Clearly, the combinatorial proofs do not respect the desiderata for proofs that we listed before. They are not step-by-step procedures, they do not involve lists of formulae and neither do they have a linguistic component. Then, seems that we should lean towards a negative answer: they are not proofs, therefore they are not syntactic.

Nevertheless, we saw above that the intuitive notion of proof does not describe all the possible formal proofs. Moreover I stated that diagrammatical formal systems generate syntactic objects, thus at least their diagrams deserve to be called proofs. Then, even if combinatorial proofs do not fall on the intuitive definition of proofs, I strongly believe these are syntactical objects. And they deserve to be called proofs.

But our question was if combinatorial proofs are syntax or semantics,

---

[6]Many other people should be thanked or blamed for the ideas that are expressed in this article. Mainly I would like to acknowledge the audience at Logica 2018 and the scholars I met at the Logic Colloquium in Udine, their comments have been crucial.

On the Nature of Combinatorial Proofs

not just simple proofs. Taking in account the previous discussion, we could summarise the differences between syntax and semantics in three major points:

1. Complexity in exponential time *vs.* Complexity in polynomial time

2. Define consequence as preservation of truth in all models. *vs.* Define consequence as the existence of some proof.

3. Gives conditions on when a claim is true in a model. *vs.* Gives rules explaining what follows from what.

We have already discussed the first point, but what about the other two? For the second point is really easy: combinatorial proofs behave exactly as any other syntactic formalism, defining consequence as the existence of that special graph-theoretical structure. We also have a soundness and completeness theorem which guarantee that saying a certain formula ϕ is true and ϕ has a combinatorial proof are equivalent[7].

The third point is what, in my opinion, makes everything harder and it is where we can notice that the line which separate these two branches is blurred.

As said earlier, I think model theoretic semantic identifies some basic notion of validity, a kind of core of the meaning of the connectives. On the other side, syntax gives the use of those symbols.

The problem of identity of proofs is generated by this wild diversification of syntax. We have a jungle of rules where they differ from one another just by some technicalities, while the essence stay intact. Think for example of operational rules for classical sequent calculus, we can have rules with some built-in notion of weakening, context-sharing or not, and so on. But can we actually say that the meaning of conjunction in LK and G3c is different?[8] I see semantics as a sort of invariant of the meaning, regardless the form of the calculus we adopted.

Combinatorial proofs glean the essential notions of connectives where, for example, a conjunction is a strong link between two formulae, and they also avoid all the syntactic technicality. In other words, combinatorial proofs represent the sharing meaning of the connectives in all the possible formalisations, and this is how they can map many proofs into the same combinatorial proof. They look at the essence, at the core of proofs and formalisms,

---

[7]Moreover, we can even add the request that the combinatorial proof is cut free.
[8]LK and G3c are two different proof systems for classical logic.

avoiding the technicalities that in this context, i.e., finding a notion of identity, are just an obstacle that hides the skeleton of proofs.

But then the question remain unanswered: are combinatorial proofs real "proofs"? As mentioned above, I believe that the two notions of semantics and syntax are not such strong dichotomy as they can looks at a first glance. Combinatorial proofs are a formalism which commits to this blended duality: we have a semantic attitude on defining connectives but we still have a very syntactical proof search. We always talk about finite structures, and this is for me the main reason why we should consider combinatorial proofs as syntactic objects. Another very important syntactic feature of combinatorial proofs is that there is a checkability criterion in polynomial time. Combinatorial proofs are constructed by a coloured graph that represents an axiomatic structure, which is built algorithmically, starting from the graph of the provable formula.

What about the explanatory power? It is undeniable that most of it is lost. What makes syntax so diversified, is that choosing a different set of rules it is possible to emphasise some property, or some structure, and understand, in the step-by-steps procedures, how the reasoning is developed. Combinatorial proofs mostly get rid of it, but they still give some informations. For example, the homomorphism embeds the logic's structural properties, and tells us which ones are involved in the proof.

I think the aim of combinatorial proofs is to emphasise the structure of formal proofs, and recognise when they are they same regardless of the formalism, and the syntactic structure. Combinatorial Proofs are semantic, because many proofs can be mapped into the same combinatorial proof; because they go for a very basic meaning of connectives, without getting involved with syntactic bureaucracy. They are semantic because they are a syntactic-free formalism, and because the get rid of the traditional step-by-step structure. But they are an undeniable syntactic object, whose meaning is given by the whole theory, and not just by the formal derivation. It still remains the possibility to consider this formalism as a sort of proof-theoretic semantics, but this would require further analysis.

References

Bundy, A., Jamnik, M., & Fugard, A. (2005). What is a proof? *Philosophical Transactions of the Royal Society of London A: Mathematical, Physical and Engineering Sciences, 363*(1835), 2377–2391.

Carbone, A. (2010). A new mapping between combinatorial proofs and sequent calculus proofs read out from logical flow graphs. *Information and Computation, 208*(5), 500–509.

Cook, S. A., & Reckhow, R. A. (1979). The relative efficiency of propositional proof systems. *The Journal of Symbolic Logic, 44*(1), 36–50.

Fomenko, A. T. (2012). *Visual Geometry and Topology*. Springer Science & Business Media.

Gamut, L. T. F. (1991). *Logic, Language, and Meaning, volume 1: Introduction to Logic* (Vol. 1). University of Chicago Press.

Gentzen, G. (1970). *The Collected Papers of Gerhard Gentzen*. North-Holland Pub. Co.

Giaquinto, M. (2008). Visualizing in mathematics. *The Philosophy of Mathematical Practice*, 22–42.

Girard, J.-Y. (1987). Linear logic. *Theoretical Computer Science, 50*(1), 1–101.

Hilbert, D. (1930). Die Grundlebung der elementahren Zahlenlehre. *Mathematische Annalen, 104*, 485–494.

Hughes, D. J. (2006a). Proofs without syntax. *Annals of Mathematics*, 1065–1076.

Hughes, D. J. (2006b). Towards Hilbert's 24th problem: Combinatorial proof invariants:(preliminary version). *Electronic Notes in Theoretical Computer Science, 165*, 37–63.

Mancosu, P. (2005). Visualization in logic and mathematics. In *Visualization, Explanation and Reasoning Styles in Mathematics* (pp. 13–30). Springer.

Straßburger, L. (2007). What is a logic, and what is a proof. *Logica Universalis*, 135–145.

Straßburger, L. (2017). Combinatorial flows and their normalisation. In *FSCD 2017-2nd International Conference on Formal Structures for Computation and Deduction* (Vol. 84, pp. 31:1–31:17).

Szabolcsi, A. (2007). *Questions About Proof Theory Vis-A-Vis Natural Language Semantics*.

Tait, W. W. (1968). Normal derivability in classical logic. In *The Syntax and Semantics of Infinitary Languages* (pp. 204–236). Springer.

Thiele, R. (2005). Hilbert and his twenty-four problems. *Mathematics and the Historian's Craft*, 243–295.

Thiele, R., & Wos, L. (2002). Hilbert's twenty-fourth problem. *Journal of Automated Reasoning, 29*(1), 67–89.

Serena Delli

Serena Delli
Universidade Nova de Lisboa
Portugal
E-mail: s.delli@campus.fct.unl.pt

A Formal Model for Explicit Knowledge as Awareness-Of Plus Awareness-That

CLAUDIA FERNÁNDEZ-FERNÁNDEZ[1] AND FERNANDO R. VELÁZQUEZ-QUESADA.

Abstract: In the context of the problem of logical omniscience, several frameworks have been proposed to model the knowledge of 'real' agents with limited reasoning abilities. One of the most important, *awareness logic*, relies on the concept of *awareness* for distinguishing what the agent 'truly' knows from what she could get if she were aware of all formulas. Still, the notion of awareness can be interpreted in different ways: it can be understood as what the agent simply entertains, without having any attitude in favour or against (*awareness of*), but also as what she has consciously recognised as true (*awareness that*). This paper proposes a formal framework that captures these two interpretations of the notion of awareness, discussing the further epistemic notions that arise from their combination (e.g., *implicit knowledge* and *explicit knowledge*) while also studying their properties and the way they interact with one another.

Keywords: awareness, explicit knowledge, awareness logic, neighbourhood semantics, dynamic epistemic logic

1 Introduction

Since the problem of logical omniscience was identified (Hintikka, 1962; see also Stalnaker, 1991), several frameworks have been proposed to model the knowledge of 'real' agents with limited reasoning abilities. To do so, knowledge has been typically split into *explicit* and *implicit*, with the former being the 'real' knowledge the agent has, and the latter being the knowledge an ideal agent would obtain.

Among the many different proposals, the *awareness logic* of Fagin and Halpern (1988) has been one of the most successful. It relies on the insight

---

[1]Partially supported by the Spanish Ministry of Science Project number TIN15-70266-C2-P-1 and the European Regional Fund Development (ERFD).

that, for an agent to know that certain φ indeed holds, it is not enough for φ to be the case in all her epistemic alternatives (as the standard epistemic logic of Hintikka, 1962 requires): the agent should be also *aware of* it. This simple but powerful idea has proven to be useful in Philosophy, Computer Science and Economics (see, e.g., the comprehensive handbook chapter by Schipper, 2015).

Still, the notion of awareness can be interpreted in different ways (cf. Dretske, 1993): it can be understood as what the agent simply entertains, without having any attitude in favour or against (*awareness-of*), but also as what she has consciously recognised as true (*awareness-that*). Thus, lacking of awareness can have different meanings. On the one hand, potential lacking of *awareness-of* yields agents who, while possibly not aware of all involved possibilities, are still ideal reasoners within the realm of what they entertain (see, e.g., the original Fagin & Halpern, 1988, and also Halpern & Rêgo, 2008; Heifetz, Meier, & Schipper, 2008). On the other hand, potential lacking of *awareness-that* yields agents which entertain all relevant possibilities, and still might not have been able to realise that a certain ψ is the case despite knowing (explicitly) both φ and $\varphi \to \psi$ (see, e.g., Konolige, 1984; Velázquez-Quesada, 2013).

Fernández-Fernández and Velázquez-Quesada (2018) proposed, at an intuitive level, a setting understanding explicit knowledge as the combination of both *awareness-of* and *awareness-that*: in order to know a given φ explicitly, the agent needs to entertain the possibility of φ, but she also needs to recognise that the formula is indeed the case.[2,3] This is a crucial feature, as considering these two kinds of awareness allows us to separate the mere fact of entertaining some information (being *aware of* φ; just a matter of attention) from recognising (acknowledging/accepting) that some φ is indeed the case (being *aware that* φ).

A setting considering both *awareness-of* and *awareness-that* gives rise to further epistemic concepts, and the diagram on Figure 1 provides a visual aid for this. While the small ellipse near the centre contains what the agent is aware that, the large dashed ellipse on the right contains what the agent is aware of. Two further 'big areas' arise: the logical consequences of what the agent is aware that (the large dotted ellipse on the left), and all truthful information the agent might become aware of (the whole domain). The former can be seen as the agent's implicit knowledge under acts of deductive

---

[2]Cf. the proposal of Grossi and Velázquez-Quesada (2015)

[3]In line with the epistemological view of knowledge as 'Justified True Belief', asking for the agent to also have a *justification* supporting her explicit knowledge.

Explicit Knowledge as Awareness-Of + Awareness-That

inference (what she would be aware-that if she were to perform all possible deductive inferences); the latter can be seen as the agent's implicit knowledge under acts of becoming aware (what she would entertain if she became aware of all relevant possibilities). Together, they define the regions **1** to **5**, described by the text next to the diagram.

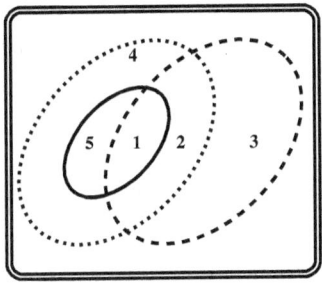

1 - What the agent is aware that and aware of (i.e., what she explicitly knows).

2 - What the agent is entertaining, has not recognised as true, but will after deductive reasoning.

3 - What the agent is entertaining, has not recognised as true, and is outside the scope of deductive reasoning.

4 - What the agent could deduce if she became aware of it.

5 - What the agent has recognised as true but is not currently entertaining.

Figure 1: Combining *awareness of* and *awareness that*.

The present proposal provides a formal logical framework capturing these intuitive notions. First, it provides a model in which *awareness that* is depicted by means of a neighbourhood function (Montague, 1970; Scott, 1970; see Pacuit, 2017 for a modern presentation), with each world's neighbourhood understood as a list containing the semantic representation of the formulas the agent has acknowledged as true. Then, *awareness of* is depicted by a single set of atomic propositions (cf. Fagin & Halpern, 1988), understood as the atoms defining the agent's current language. Explicit knowledge is then defined as what the agent is both aware of and aware that, with implicit knowledge defined as what the agent would know explicitly after performing every possible deductive inference. The proposal also provides a formal language for describing such structures, allowing a formal discussion of the concepts' properties and subtle interactions. The main aim is not only to provide a slightly different approach for modelling the knowledge of more 'real' agents, but also to shed light to other epistemic concepts that arise when combining these two types of awareness.

Claudia Fernández-Fernández and Fernando R. Velázquez-Quesada

2 The formal framework

Let P be a non-empty enumerable set of atomic propositions.

Definition 1 (Awareness neighbourhood model) *An* awareness neighbourhood *model (ANM) is a tuple* $M = \langle W, N, V, A \rangle$ *where (i)* W, *sometimes denoted as* $\mathcal{D}_M$, *is a non-empty set (whose elements are called* possible worlds*), (ii)* $N : W \to \wp(\wp(W))$ *is a* neighbourhood function *(assigning a set of sets of worlds to each possible world, with $N(w)$ called the* neighbourhood of w*), (iii)* $V : \text{P} \to \wp(W)$ *is a* valuation function *(indicating the set of possible worlds in which each atom is true), and (iv)* $A \subseteq \text{P}$ *is the* atomic awareness set *(indicating the set of atoms the agent is aware of).*

Note that *awareness-that* corresponds to the (local) neighbourhood function N and *awareness-of* is generated by the (global) atomic set A.

Definition 2 (Language) *An ANM is described by the language* $\mathcal{L}$, *whose formulas* φ, ψ *are given by*

$$\varphi, \psi ::= \top \mid p \mid \neg\varphi \mid \varphi \wedge \psi \mid \text{A}^\text{t}\,\varphi \mid \text{A}^\circ\,\varphi \mid [*]\,\varphi,$$

with $p \in \text{P}$. *Formulas of the form* $\text{A}^\text{t}\,\varphi$ *are read as "the agent is aware that* φ*", those of the form* $\text{A}^\circ\,\varphi$ *are read as "the agent is aware of* φ*", and* $[*]\,\varphi$ *expresses that "after the agent performs every possible deductive inference, φ holds". The set of atoms of any given* $\varphi \in \mathcal{L}$, *denoted by* $\text{atm}(\varphi)$, *is defined in the standard way.*[4]

Definition 3 (Semantic interpretation) *The function* $\llbracket \cdot \rrbracket^M : \mathcal{L} \to \wp(W)$, *returning the set of worlds in a given ANM* $M = \langle W, N, V, A \rangle$ *in which a given* φ *is the case (the* truth-set *of* φ *in* M*), is defined inductively as follows. The cases for* $\top$, *atoms and Boolean operators are standard:*

$$\llbracket \top \rrbracket^M := W,\ \llbracket p \rrbracket^M := V(p),\ \llbracket \neg\varphi \rrbracket^M := W \setminus \llbracket \varphi \rrbracket^M,\ \llbracket \varphi \wedge \psi \rrbracket^M := \llbracket \varphi \rrbracket^M \cap \llbracket \psi \rrbracket^M.$$

For $\text{A}^\circ\,\varphi$, *the formula is either globally true (when the agent is aware of all φ's atoms) or else globally false (otherwise):*

$$\llbracket \text{A}^\circ\,\varphi \rrbracket^M := \begin{cases} W & \text{if } \text{atm}(\varphi) \subseteq A; \\ \varnothing & \text{otherwise.} \end{cases}$$

---

[4]That is, $\text{atm}(\top) := \varnothing$, $\text{atm}(p) := \{p\}$, $\text{atm}(\neg\varphi) := \text{atm}(\varphi)$, $\text{atm}(\varphi \wedge \psi) := \text{atm}(\varphi) \cup \text{atm}(\psi)$, $\text{atm}(\text{A}^\text{t}\,\varphi) := \text{atm}(\varphi)$, $\text{atm}(\text{A}^\circ\,\varphi) := \text{atm}(\varphi)$, $\text{atm}([*]\,\varphi) := \text{atm}(\varphi)$.

For $A^t \varphi$, the formula is true at a world w in M if and only if the truth-set of φ in M is in the neighbourhood of w:

$$[\![A^t \varphi]\!]^M := \left\{ w \in W \mid [\![\varphi]\!]^M \in N(w) \right\}.$$

The remaining modality, $[*]$, is semantically interpreted not over M, but rather over its augmentation: *the model M^* that results from making the neighbourhood of each world a set that contains the neighbourhood's core $\bigcap N(w)$ and is closed under supersets.[5] More precisely, from the given $M = \langle W, N, V, A \rangle$, define $M^* = \langle W, N^*, V, A \rangle$ with N^* such that*

$$N^*(w) := \{ U \subseteq W \mid \bigcap N(w) \subseteq U \}.$$

Then,
$$[\![[*]\varphi]\!]^M := [\![\varphi]\!]^{M^*}.$$

As it will be recalled (subsection 3.2), in M^* the modality A^t behaves as $\Box$ does in relational models; this is the reason behind the intuitive reading of formulas of the form $[*]\varphi$, and the reason why $[*]$ is called here the deductive closure *modality.*

Satisfiability and validity are defined in the standard way, with the latter denoted as usual ($\Vdash \varphi$).

3 Concepts, their properties and their relationship

The formal framework allows us to provide formal definitions for the notions the diagram on Figure 1 sketches.[6] Besides *awareness-of* and *awareness-that*, the concepts that will be discussed in detail are the following two, both falling under what the agent entertains:

- **Explicit knowledge:** $K_{Ex}\varphi := A^\circ \varphi \wedge A^t \varphi$, what the agent is entertaining (is aware of) and has acknowledged as true (is aware that); this is what she 'really' knows.

- **Implicit knowledge:** $K_{Im}\varphi := A^\circ \varphi \wedge [*]A^t \varphi$, what the agent is entertaining (is aware of) and will acknowledge as true after applying all possible deductive inferences.

---
[5] For specific details see, e.g., (Chellas, 1980, Section 7.3)
[6] For a more detailed explanation of the theoretical framework see (Fernández-Fernández & Velázquez-Quesada, 2018).

Claudia Fernández-Fernández and Fernando R. Velázquez-Quesada

Note how one can also define further epistemic concepts, as the respective *awareness-of-less* counterparts of the previous two. What the agent has acknowledged as true but is not currently entertaining (what will become explicitly known after she becomes aware of it), $K_{Ex}^{-A^\circ} \varphi := \neg A^\circ \varphi \wedge A^t \varphi$, can be called *'disassociated' knowledge*; what she is not currently entertaining, and yet she can deduce from what she has acknowledged as true (i.e., what she could deduce after becoming aware of what she is aware that), $K_{Im}^{-A^\circ} \varphi := \neg A^\circ \varphi \wedge [*] A^t \varphi$, can be called *currently 'unreachable' knowledge*.

3.1 Basic properties and relationships

It is now time to discuss the properties of the crucial concepts, and how they are related to their definitions elsewhere in the literature.

Awareness-of. The concept of *awareness-of*, A°, is understood here as what the agent *entertains*. Thus, *awareness-of* is a matter of attention, and by itself it does not imply any attitude pro or con. Here, A° is defined in terms of a global set of atomic propositions, the atomic awareness set A. As a consequence of this, *the agent is aware of the concept of 'truth'*: $\Vdash A^\circ \top$. This does not say that she is aware that this 'truth' holds everywhere; it simply states that she 'knows' that such concept exists. Technically, this is the case because $\top$, a primitive in the language[7], does not contain any atomic proposition.

Still, despite being aware of the concept of truth, the agent does not need to be aware of formulas that are always true: $\Vdash \varphi$ *does not imply* $\Vdash A^\circ \varphi$, and thus there might be validities the agent does not entertain. Technically, this is the case because a validity might have atoms (e.g., $p \vee \neg p$), and no atom is required to be in the atomic awareness set (e.g., p does not need to be in A).

On the other hand, *awareness-of* is defined not in terms of a set of formulas (as in the logic of general awareness by Fagin and Halpern) but rather in terms of a set of *atomic propositions*. As discussed already by Fagin and Halpern (1988), this makes the concept of *awareness-of* closed under *subformulas and superformulas*. More precisely,

$\Vdash A^\circ \neg \varphi \leftrightarrow A^\circ \varphi,$ $\qquad\qquad \Vdash A^\circ A^\circ \varphi \leftrightarrow A^\circ \varphi,$
$\Vdash A^\circ (\varphi \wedge \psi) \leftrightarrow (A^\circ \varphi \wedge A^\circ \psi),$ $\qquad \Vdash A^\circ A^t \varphi \leftrightarrow A^\circ \varphi,$
$\qquad\qquad\qquad\qquad\qquad\qquad\qquad\qquad \Vdash A^\circ [*] \varphi \leftrightarrow A^\circ \varphi.$

---

[7]Not defined by an abbreviation of the form $p \vee \neg p$, as done in other proposals.

Explicit Knowledge as Awareness-Of + Awareness-That

In particular, note what the formulas on the right column indicate. The first states that awareness of awareness-of a formula is equivalent to awareness of the formula. The second and the third indicate, with some paraphrasing, that entertaining φ is equivalent to entertaining the possibility of having accepted φ as true (the second) and also equivalent to entertaining the possibility of φ being true after deductive inference (the third).

An important difference between the *awareness-of* discussed here and the one by Fagin and Halpern (1988) is that, while the latter is local (what belongs to the awareness set assigned to the given evaluation point), the one proposed here is global (what belongs to the atomic awareness set assigned to *the whole model*). An intermediate position is held by Grossi and Velázquez-Quesada (2015), where (atom-based) *awareness-of* is defined as those formulas whose atoms appear in the awareness set of all worlds the agent cannot distinguish from the current evaluation point. Such definition makes sense in the multi-agent setting that the referred paper studies; in the single-agent case examined here, one can simply assume that the worlds in the model are exactly those that are relevant for the agent under discussion, thus making the two definitions conceptually equivalent.

Awareness-that. The concept of *awareness-that*, A^t, is understood here as what the agent has accepted/acknowledged as true; in this sense, it is a form of 'explicit' information. Still, it is not called *explicit knowledge* because the agent might not entertain such piece of information at the current stage. Thus, even though *awareness-that* is acknowledgement of truth, acceptance by itself does not imply that the agent is still entertaining such piece of information (she might have moved on to a different topic), and therefore it does not imply explicit knowledge either.

Here, A^t is defined as what appears in the neighbourhood of the current evaluation point.[8] In fact, the neighbourhood of a given world, a set of sets of worlds, can be understood as the list of formulas the agent has acknowledged as true, the important point being that these formulas are not represented syntactically (as a string of symbols), but rather semantically (as the set of worlds in the model in which the formula is true). Because of this purely semantic representation, the concept of *awareness-that* has an important closure property: $\Vdash \varphi \leftrightarrow \psi$ implies $\Vdash A^t \varphi \leftrightarrow A^t \psi$. In other words,

---

[8]In fact, A^t's semantic interpretation is the 'set' semantic interpretation of the $\Box$-operator in neighbourhood models. (Recall: the 'subset' semantic interpretation makes $\Box \varphi$ true at w in M not only when $[\![\varphi]\!]^M$ is in $N(w)$, but also when any of its *subsets* is: $[\![\Box \varphi]\!]^M := \{ w \in W \mid \text{there is } U \in N(w) \text{ such that } U \subseteq [\![\varphi]\!]^M \}$.)

the agent's *awareness-that* is *closed under logical equivalence*.[9] Thus, the agent is indeed omniscient in some sense as, at the level of acknowledgement, she cannot tell apart formulas that are true in exactly the same worlds in all models.[10]

Note that closure under logical equivalence is the only closure property the notion of *awareness-that* has. Different from other semantic representations of information, as the $\Box$-operator in standard epistemic logic under relational models, *(i)* $\Vdash \varphi$ does not imply $\Vdash A^t \varphi$, *(ii)* $\not\Vdash (A^t \varphi \wedge A^t \psi) \to A^t(\varphi \wedge \psi)$, *(iii)* $\not\Vdash A^t(\varphi \wedge \psi) \to A^t \varphi$ and $\not\Vdash A^t(\varphi \wedge \psi) \to A^t \psi$. The reason for the failure of these properties is that no neighbourhood needs to have any closure property. In particular, *(i)* none of them needs to contain the whole domain (so $\mathcal{D}_M$, the truth-set of any validity, does not need to be in $N(w)$), *(ii)* none of them needs to be closed under intersections (so $[\![\varphi]\!]^M, [\![\psi]\!]^M \in N(w)$ does not imply $[\![\varphi]\!]^M \cap [\![\psi]\!]^M = [\![\varphi \wedge \psi]\!]^M \in N(w)$), and *(iii)* none of them needs to be closed under supersets (so $[\![\varphi \wedge \psi]\!]^M = [\![\varphi]\!]^M \cap [\![\psi]\!]^M \in N(w)$ implies neither $[\![\varphi]\!]^M \in N(w)$ nor $[\![\psi]\!]^M \in N(w)$). As a consequence of the latter two, *awareness-that* is not closed under modus ponens: $\not\Vdash A^t(\varphi \to \psi) \to (A^t \varphi \to A^t \psi)$.[11]

Awareness-of and *awareness-that*. A property relating A° and A^t has been discussed already ($\Vdash A^\circ A^t \varphi \leftrightarrow A^\circ \varphi$). Yet, for readers familiarised with awareness logic, the fact that *awareness-of* is global might suggest a further relationship between the two concepts: that the agent 'knows her own awareness'. Indeed, in the original work of Fagin and Halpern (1988), the fact that the awareness set of all worlds in the model is the same implies not only $\Vdash A\varphi \to \Box A\varphi$ (if the agent is aware of φ, then she knows this) but also $\Vdash \neg A\varphi \to \Box \neg A\varphi$ (if she is not aware of φ, then she knows this).

In the present setting, analogous properties do not need to hold: the agent does not need to acknowledge any formula, and thus she needs to acknowledge neither her awareness, $\not\Vdash A^\circ \varphi \to A^t A^\circ \varphi$, nor her unawareness, $\not\Vdash \neg A^\circ \varphi \to A^t \neg A^\circ \varphi$.

---

[9]But, as it will be discussed later, this does not mean that the agent's *explicit knowledge* is closed under logical equivalence.

[10]This concept of *awareness that* (sometimes called *explicit knowledge*) lacks this property in other proposals (Grossi & Velázquez-Quesada, 2015; Konolige, 1984) simply because they represent it as a set of formulas, and such set is not required to have any closure property.

[11]In this semantic setting, a modus ponens is a sequence of three steps: conjunction introduction to go from $[\![p]\!]^M$ and $[\![p \to q]\!]^M$ to $[\![p \wedge (p \to q)]\!]^M$, logical equivalence to go from the latter to $[\![p \wedge q]\!]^M$, and conjunction elimination to go from the latter to $[\![q]\!]^M$.

Explicit Knowledge as Awareness-Of + Awareness-That

Explicit knowledge. The concept of *explicit knowledge*, K_{Ex}, is understood as those pieces of information the agent has acknowledged as true and is currently entertaining. Thus, explicit knowledge is what the agent is aware-of and aware-that: $K_{Ex}\, \varphi := A^\circ \varphi \wedge A^t \varphi$. Which are the consequences of this definition?

About *validities*. The agent does not need to know explicitly any validity, and none of the awareness concepts is, by itself, enough to guarantee that a validity is explicitly known: $\Vdash \varphi$ implies neither $\Vdash K_{Ex}\,\varphi$, nor $\Vdash A^\circ \varphi \to K_{Ex}\,\varphi$, nor $\Vdash A^t \varphi \to K_{Ex}\,\varphi$.

About *logical equivalence*. Even though *awareness-that* is closed under such property, *explicit knowledge* does not need to: $\Vdash \varphi \leftrightarrow \psi$ does not imply $\Vdash K_{Ex}\,\varphi \leftrightarrow K_{Ex}\,\psi$ as, even though A^t is closed under logical equivalence, A° is not. Still, *awareness-of* is the only piece that is missing. Thus, if two formulas are logically equivalent and the agent is aware of the second, explicit knowledge of the first implies explicit knowledge of the latter: $\Vdash \varphi \leftrightarrow \psi$ implies $\Vdash (K_{Ex}\,\varphi \wedge A^\circ \psi) \to K_{Ex}\,\psi$.

About *closure under modus ponens*. Awareness-that lacks this property, and thus explicit knowledge lacks it too: $\not\Vdash K_{Ex}(\varphi \to \psi) \to (K_{Ex}\,\varphi \to K_{Ex}\,\psi)$. This is already shared by the explicit knowledge of Fagin and Halpern (1988) (which, recall, is defined as implicit knowledge, $\Box \varphi$, plus *awareness-of*, $A\,\varphi$). However, different from Fagin and Halpern's proposal, here being aware of the consequent does not give the agent explicit knowledge about it: $\not\Vdash K_{Ex}(\varphi \to \psi) \to ((K_{Ex}\,\varphi \wedge A^\circ \psi) \to K_{Ex}\,\psi)$. The agent might know explicitly an implication and its antecedent, and she might be entertaining the consequent, but still might fail to have explicit knowledge of the latter. What the agent is missing is realising that the consequent is actually the case:[12] $\Vdash K_{Ex}(\varphi \to \psi) \to ((K_{Ex}\,\varphi \wedge A^t \psi) \to K_{Ex}\,\psi)$.

3.2 Effects of the *augmentation* operation

The augmentation operation makes the neighbourhood of each world a set that contains the neighbourhood's core and is closed under supersets. Equivalently, one can understand the neighbourhood of each world after the augmentation operation as the result of adding the full domain to the original neighbourhood, and then close it under supersets and arbitrary intersections. It is well-known (e.g., Chellas, 1980, Theorem 7.9) that in the resulting model, the *augmented* model M^*, the operator A^t behaves as the standard $\Box$ does in relational models. Hence, occurrences of A^t under the

---

[12]In this, the present setting coincides with Grossi and Velázquez-Quesada (2015).

scope of the modality $[*]$ can be understood as what the agent will acknowledge as true after she applies every possible deductive inference, making her *awareness-that* closed under logical consequence.[13] In this sense, $(\cdot)^*$ can be understood as a *full deductive inference* operation over A^t.

Awareness-that after deductive inference. The operation $(\cdot)^*$ makes the agent's *awareness-that* closed under logical consequence; thus, after it, the agent acknowledges every validity: $\Vdash \varphi$ implies $\Vdash [*]\, A^t\, \varphi$. Moreover, after the operation, what the agent acknowledges is closed under both conjunction introduction, $\Vdash [*]((A^t\,\varphi \wedge A^t\,\psi) \to A^t(\varphi \wedge \psi))$, and conjunction elimination, $\Vdash [*](A^t(\varphi \wedge \psi) \to A^t\,\varphi)$ and $\Vdash [*](A^t(\varphi \wedge \psi) \to A^t\,\psi)$. From the last two properties and the previous closure under logical equivalence, it follows that *awareness-that* is closed under modus ponens, $\Vdash [*](A^t(\varphi \to \psi) \to (A^t\,\varphi \to A^t\,\psi))$. The operation is a total function, so the last three properties can be described as $\Vdash ([*]\,A^t\,\varphi \wedge [*]\,A^t\,\psi) \to [*]\,A^t(\varphi \wedge \psi)$, both $\Vdash [*]\,A^t(\varphi \wedge \psi) \to [*]\,A^t\,\varphi$ and $\Vdash [*]\,A^t(\varphi \wedge \psi) \to [*]\,A^t\,\psi$, and $\Vdash [*]\,A^t(\varphi \to \psi) \to ([*]\,A^t\,\varphi \to [*]\,A^t\,\psi)$. Thus, the closure operation indeed makes A^t behave as $\Box$ under relational models.

Even more: after the operation the agent acknowledges her own *awareness-of*: $\Vdash [*](A^\circ\,\varphi \to A^t\,A^\circ\,\varphi)$ and $\Vdash [*](\neg A^\circ\,\varphi \to A^t\,\neg A^\circ\,\varphi)$ or, since the operation does not affect awareness sets, $\Vdash A^\circ\,\varphi \to [*]\,A^t\,A^\circ\,\varphi$ and $\Vdash \neg A^\circ\,\varphi \to [*]\,A^t\,\neg A^\circ\,\varphi$. Hence, the operation makes the relationship between A° and A^t as it is in standard awareness models when the notion of *awareness-of* is global.

Implicit knowledge. This concept is defined here as what the agent currently entertains and will recognize as true *after* performing all possible deductive inferences, $K_{Im}\,\varphi := A^\circ\,\varphi \wedge [*]\,A^t\,\varphi$.

For its properties, note how the agent does not need to know implicitly every validity, $\Vdash \varphi$ does not imply $\Vdash K_{Im}\,\varphi$, the reason being that she might be unaware of (some of) the involved atoms.[14] Nevertheless, different from the *explicit knowledge* case, *awareness-of* is enough; the agent knows

---

[13]Velázquez-Quesada (2013) already explores this idea of using the known relationship between neighbourhood models and relational models to contrast knowledge before deductive reasoning with knowledge after. Still, the technical details are slightly different, as in the finite case that the referred paper studies, the augmented model is (equivalently) defined as the result of adding the full domain to each neighbourhood, and then closing it under *finite* intersections and supersets. Also, the referred paper's notion of explicit knowledge is different from the current proposal, as it does not take the concept of *awareness-of* into account.

[14]This is different from what happens with the same notion in the work of Fagin and Halpern (1988), where the agent knows implicitly every validity.

Explicit Knowledge as Awareness-Of + Awareness-That

implicitly any validity she is currently entertaining: $\Vdash \varphi$ implies $\Vdash A^\circ \varphi \to K_{Im}\varphi$.

The agent's implicit knowledge is *not closed under logical equivalence*, $\Vdash \varphi \leftrightarrow \psi$ does not imply $\Vdash K_{Im}\varphi \leftrightarrow K_{Im}\psi$, the *only* reason being, again, that the *awareness-of* requirement might fail (thus, $\Vdash \varphi \leftrightarrow \psi$ implies $\Vdash (K_{Im}\varphi \wedge A^\circ \psi) \to K_{Im}\psi$). On the other hand, implicit knowledge is *closed under conjunction introduction* ($\Vdash (K_{Im}\varphi \wedge K_{Im}\psi) \to K_{Im}(\varphi \wedge \psi)$) since, *(i)* after the operation, awareness-that has such property (here its 'distributed' version is the useful one), and *(ii)* awareness-of is closed under superformulas. It is also closed under *conjunction elimination* ($\Vdash K_{Im}(\varphi \wedge \psi) \to K_{Im}\varphi$ and $\Vdash K_{Im}(\varphi \wedge \psi) \to K_{Im}\psi$) as *(i)* after the operation, *awareness-that* has such property (again, its 'distributed' version is the useful one), and *(ii) awareness-of* is closed under subformulas. These properties, together with the fact that *awareness-of* is closed under subformulas, tell us that implicit knowledge is *closed under modus ponens*: $\Vdash K_{Im}(\varphi \to \psi) \to (K_{Im}\varphi \to K_{Im}\psi)$.

Notice how, while the implicit knowledge of Fagin and Halpern (1988) contains all validities and is closed under logical consequence, the implicit knowledge as defined here might not contain all validities and, despite being closed under modus ponens, it does not need to be closed under logical equivalence. This is because, while implicit knowledge of Fagin and Halpern (1988) is the agent's 'semantic' information (given by the modal operator $\square$), here it is the closure under modus ponens of what the agent has acknowledged as true ($[*] A^t \varphi$) and is currently entertaining ($A^\circ \varphi$); it is the closure under modus ponens of what the agent knows explicitly. This highlights the crucial difference between the understanding of explicit knowledge in both settings. What is needed for implicit knowledge of Fagin and Halpern (1988) to be explicit is for the agent to be aware of the given formula. However, in this proposal, knowing a formula implicitly already makes the agent aware of it. Thus, here, what is needed for implicit knowledge to become explicit is not an act of awareness raising; what is needed is rather an act of *deductive inference*.

Moorean phenomena. In proposals dealing with implicit and explicit knowledge, a particular property is recurrent: explicit knowledge is also implicit knowledge. In the work of Fagin and Halpern (1988), this follows from the fact that explicit knowledge is defined as the implicit knowledge that satisfies an additional requirement (*awareness-of*); in settings distinguishing explicit information from implicit one by means of deductive reasoning (e.g., Konolige, 1986), this follows from the fact that deductive inference is monotone.

This property, seemingly not only natural but rather essential, is not satisfied by the presented setting: $\not\vdash K_{Ex}\,\varphi \to K_{Im}\,\varphi$. The reason is, as shown by Velázquez-Quesada (2013), that what the agent has acknowledged as true at some particular stage does not need to be acknowledged as true after the augmentation operation, that is,

Fact 1 $\not\vdash A^t\,\varphi \to [*]\,A^t\,\varphi$

Proof. Assume that $\varphi := \neg A^t\,q$ and consider the four-worlds model $M = \langle W = \{w_1, w_2, w_3, w_4\}, N, V, \varnothing \rangle$ over the set of atomic propositions $\{p, q\}$, with $V(p) = \{w_1, w_2\}$ and $V(q) = \{w_1, w_3\}$. The *awareness-of* set A is not relevant (hence empty, for simplicity), but the neighbourhood function is: suppose it is given by $N(w_1) := \{\{w_1, w_2\}, \{w_1, w_3, w_4\}, W\}$ and $N(w_2) = N(w_3) = N(w_4) := \varnothing$. Since $[\![\neg A^t\,q]\!]^M = W$ is in $N(w_1)$, we have $w_1 \in [\![A^t\,\neg A^t\,q]\!]^M$.

However, the neighbourhood function of the augmented model, N^*, is such that $N^*(w_1) = \{U \subseteq W \mid \{w_1\} \subseteq U\}$ and $N^*(w_2) = N^*(w_3) = N^*(w_4) = \{W\}$. Note how $[\![\neg A^t\,q]\!]^{M^*} = \{w_2, w_3, w_4\}$ is *not* in $N^*(w_1)$, so $w_1 \notin [\![A^t\,\neg A^t\,q]\!]^{M^*}$, that is, $w_1 \notin [\![[*]\,A^t\,\neg A^t\,q]\!]^M$. □

Thus, while in M the agent has acknowledged $\neg A^t\,q$ as true at w_1 (i.e., $w_1 \in [\![A^t\,\neg A^t\,q]\!]^M$), the operation changes this: in M^* the agent has not acknowledged $\neg A^t\,q$ as true at w_1 (i.e., $w_1 \notin [\![A^t\,\neg A^t\,q]\!]^{M^*}$). One just needs to make the agent aware of the involved formula $\neg A^t\,q$ (e.g., take $A := \{p, q\}$) to obtain a model (M) and a world (w_1) in which the agent knows a formula ($\neg A^t\,q$) explicitly,

$$w_1 \in [\![K_{Ex}\,\neg A^t\,q]\!]^M = [\![A^\circ\,\neg A^t\,q \wedge A^t\,\neg A^t\,q]\!]^M$$
$$= [\![A^\circ\,\neg A^t\,q]\!]^M \cap [\![A^t\,\neg A^t\,q]\!]^M,$$

and yet she does not know it implicitly,

$$w_1 \notin [\![K_{Im}\,\neg A^t\,q]\!]^M = [\![A^\circ\,\neg A^t\,q \wedge [*]\,A^t\,\neg A^t\,q]\!]^M$$
$$= [\![A^\circ\,\neg A^t\,q]\!]^M \cap [\![[*]\,A^t\,\neg A^t\,q]\!]^M.$$

So, is there some fundamental problem with the current proposal? To answer this, first note how the 'explicit is implicit' property does not always fail. In fact, it holds for a large class of formulas, including not only the

Explicit Knowledge as Awareness-Of + Awareness-That

purely propositional ones, but also all those whose truth-set *does not shrink* as a consequence of the augmentation operation:

Proposition 1 $\Vdash \varphi \to [*]\,\varphi$ *implies* $\Vdash \mathrm{K}_{Ex}\,\varphi \to \mathrm{K}_{Im}\,\varphi$.

Proof. See that of Velázquez-Quesada (2013, Proposition 2). □

Then, why do the rest of the formulas fail? The provided counterexample, $\neg \mathrm{A}^t q$, shows one of their crucial feature: they express not ontic facts, but rather *epistemic* situations and, in particular *negative awareness-that* situations. Indeed, $\neg \mathrm{A}^t q$ expresses that the agent has not acknowledged q as true, and then $\mathrm{A}^t \neg \mathrm{A}^t q$ says that the agent has acknowledged this. In other words, and considering her *awareness-of*, the agent knows explicitly that she does not know q explicitly.

However, the agent might have enough information to get to know what she currently knows she does not have. Indeed, in the provided model (and with $A := \{p, q\}$), she knows explicitly both p and $p \to q$ (both $[\![p]\!]^M = \{w_1, w_2\}$ and $[\![p \to q]\!]^M = \{w_1, w_3, w_4\}$ are in $N(w_1)$). Then, after deductive reasoning, she will realise that q is indeed the case, hence knowing q explicitly; but then, she will automatically stop acknowledging (and thus stop knowing explicitly) that she did not know q explicitly. In other words, she might know explicitly that she does not know q, but such high-order knowledge will be gone once she gets to know that q is indeed the case.

The reason for the failure of the 'explicit is implicit' property is that the agent has knowledge not only about propositional facts but also about her own (and eventually other agents') knowledge. This knowledge (semantically, the neighbourhood function) changes through the closure operation, so the agent might know something explicitly at some point, and yet not know it explicitly after (semantically, the *awareness-that* component is the key: we might have $U \in N(w)$ with $U = [\![\varphi]\!]^M$ for some φ, but even though this implies $U \in N^*(w)$, nothing guarantees $U = [\![\varphi]\!]^{M^*}$). This is an instance of the so called 'Moorean phenomena', which occurs when an epistemic action invalidates itself. In its best known incarnation, this phenomenon appears as formulas that become false after being truthfully announced (Holliday & Icard III, 2010; van Ditmarsch & Kooi, 2006); here, it appears as formulas that become false after deductive inference.

4 Summary and ongoing work

We have seen so far our model for representing the concept of *explicit knowledge*, based on the combination of *awareness that* and *awareness of*. Afterwards we have shown some properties of these concepts that highlight not only the relationship between them, but also some of the advantages our model may have with respect to alternative proposals such as the logic of awareness (Fagin & Halpern, 1988) or deduction systems (Konolige, 1984).

What our system is still missing is a further exploration into the dynamics, i.e., we need to define further epistemic actions that will show how the information changes throughout the process of gaining or loosing knowledge. Actions like becoming (un)aware, performing a single-step deductive inference or observing a piece of information will be crucial for this. We would also like to complete a comprehensive comparison for the notions of *awareness of* and *awareness that* with similar proposals, such as the concept of *topics* (Berto & Hawke, 2018) for the former, or the proposals of *explicit knowledge* that do not incorporate the awareness of the agent for the latter (Artemov & Nogina, 2005; Konolige, 1984; Velázquez-Quesada, 2013).

For future work we leave the axiomatization of the framework and its epistemic actions, together with the modelling of the intuitive interpretation of the concept of justification, understood here as the actions that turn information into explicit knowledge.

References

Artemov, S. N., & Nogina, E. (2005). Introducing justification into epistemic logic. *Journal of Logic and Computation, 15*(6), 1059–1073.

Berto, F., & Hawke, P. (2018). Knowability relative to information. *Mind*.

Chellas, B. F. (1980). *Modal Logic: An Introduction*. Cambridge.: Cambridge University Press.

Dretske, F. (1993). Conscious experience. *Mind, 102*(406), 263–283.

Fagin, R., & Halpern, J. Y. (1988). Belief, awareness, and limited reasoning. *Artificial Intelligence, 34*(1), 39–76.

Fernández-Fernández, C., & Velázquez-Quesada, F. R. (2018). Reconsidering the 'ingredients' of explicit knowledge. In P. Arazim & T. Lávička (Eds.), *The Logica Yearbook 2017* (pp. 47–60). College Publications.

Grossi, D., & Velázquez-Quesada, F. R. (2015). Syntactic awareness in logical dynamics. *Synthese, 192*(12), 4071–4105.

Halpern, J. Y., & Rêgo, L. C. (2008). Interactive unawareness revisited. *Games and Economic Behavior, 62*(1), 232–262.

Heifetz, A., Meier, M., & Schipper, B. C. (2008). A canonical model for interactive unawareness. *Games and Economic Behavior*, 62(1), 304–324.
Hintikka, J. (1962). *Knowledge and Belief*. Ithaca: Cornell U.P.
Holliday, W. H., & Icard III, T. F. (2010). Moorean phenomena in epistemic logic. In L. D. Beklemishev, V. Goranko, & V. B. Shehtman (Eds.), *Advances in Modal Logic 8* (pp. 178–199). College Publications.
Konolige, K. (1984). *A Deduction Model of Belief and its Logics* (Unpublished doctoral dissertation). Stanford University.
Konolige, K. (1986). What awareness isn't: A sentential view of implicit and explicit belief. In J. Y. Halpern (Ed.), *Proceedings of the 1st Conference on Theoretical Aspects of Reasoning about Knowledge* (pp. 241–250). Morgan Kaufmann.
Montague, R. (1970). Universal grammar. *Theoria*, 36(3), 373–398.
Pacuit, E. (2017). *Neighborhood Semantics for Modal Logic*. Springer.
Schipper, B. C. (2015). Awareness. In H. van Ditmarsch, J. Y. Halpern, W. van der Hoek, & B. Kooi (Eds.), *Handbook of Epistemic Logic*. London: College Publications.
Scott, D. (1970). Advice on modal logic. In K. Lambert (Ed.), *Philosophical Problems in Logic* (pp. 143–173). Dordrecht: Reidel.
Stalnaker, R. (1991). The problem of logical omniscience, I. *Synthese*, 89(3), 425–440.
van Ditmarsch, H. P., & Kooi, B. P. (2006). The secret of my success. *Synthese*, 151(2), 201–232.
Velázquez-Quesada, F. R. (2013). Explicit and implicit knowledge in neighbourhood models. In D. Grossi, O. Roy, & H. Huang (Eds.), *Logic, Rationality, and Interaction* (Vol. 8196, pp. 239–252). Springer.

Claudia Fernández-Fernández
University of Málaga
Spain
E-mail: cffernandez@uma.es

Fernando R. Velázquez-Quesada
University of Amsterdam
Institute for Logic, Language and Computation
The Netherlands
E-mail: F.R.VelazquezQuesada@uva.nl

Faithful Relative Truth Definability and a Truth-Theoretic Strength of Type-Free Truth

DAICHI HAYASHI[1]

Abstract: In this paper, we consider the proof-theoretic relationship between some typed and type-free theories of truth. For this purpose, we propose faithful relative truth definability, a notion stricter than relative truth definability given by Fujimoto (2010). Then we show that some type-free theories satisfy faithful relative truth definability of the corresponding typed theories respectively, and claim that these type-free theories can be regarded as non-problematic extensions of typed ones.

Keywords: axiomatic theories of truth, proof-theoretic strength, faithful interpretation

1 Introduction

As well known, we cannot in an usual arithmetic, say, in Peano Arithmetic (PA) define the truth predicate Tx satisfying the T-schema:

$$T\ulcorner\varphi\urcorner \leftrightarrow \varphi,$$

where φ is any sentence of the language of PA and $\ulcorner\varphi\urcorner$ is its name. Thus, if we adopt classical logic, then we need to restrict this T-schema in order to obtain a consistent theory of truth.

The ramified theory $RT_{<1}$ of truth characterizes truth in PA, and thus restricts the T-schema to arithmetical sentences. Next, we can think of truth in $RT_{<1}$, which is characterized using a new truth predicate, and we call this new theory $RT_{<2}$. We can iterate this process transfinitely and obtain the ramified theories $RT_{<\alpha}$. In contrast, type-free theories keep the language, but restrict the range of sentences to which T-schema applies. For example, the Kripke-Feferman theory KF given by Feferman (1991) proves the

---

[1] I would like to show my greatest appreciation to Prof. Katsuhiko Sano whose comments and suggestions on the draft were innumerably valuable throughout the course of my study.

T-schema for all sentences in which truth predicates occur only positively. Another famous example is the Friedman-Sheard theory FS given by Friedman and Sheard (1987) which cannot in general prove $T\ulcorner T(t)\urcorner \leftrightarrow T(t)$. In addition to these different approaches to truth, various theories of truth have been proposed by many authors (see Halbach, 2014). Therefore, it is important to give the means to compare these theories of truth for the unified understanding of truth. Especially, Fujimoto (2010) proposed relative truth definability as a criteria to compare theories of truth. If a theory S is relatively truth definable in the other theory T, then T can be seen as involving all theorems of S via this translation. In this sense, relative truth definability is related to the truth-theoretic strength of a given theory of truth.

Then, can we say more about how much T is stronger than S? The theory T may have almost the same strength as S, or may be problematically much stronger than S and hence the acceptance of S may not imply that of T. In this paper we propose one criteria, faithful relative truth definability, for T being regarded as a non-problematic extension of S.

The structure of this paper is as follows. In section 2 we give some technical preliminaries. Next, we introduce Fujimoto's relative truth definition and give a definition of faithful relative truth definition in section 3. In sections 4 and 5 we show that the type-free theories FS and KF satisfy faithful relative truth definability of the corresponding ramified theories $RT_{<\omega}$ and $RT_{<\varepsilon_0}$ respectively, and we finally claim that these type-free systems are non-problematic extensions of typed ones.

2 Technical preliminaries

Let the language $\mathcal{L}$ of PA have $\vee, \forall, \neg$ and $=$ as logical symbols. The other logical symbols $\wedge, \exists, \rightarrow$ and $\leftrightarrow$ are defined in a standard way. The remaining $\mathcal{L}$-symbols are the constant symbol $\bar{0}$ and the function symbols for corresponding primitive recursive functions. For each $n \in \omega$, we denote its numeral by $\bar{n}$.

For a given language $\mathcal{L}'$, we fix its Gödel numbering, and use the notation $\ulcorner e \urcorner$ for (the numeral of) the Gödel number of an $\mathcal{L}'$-expression e.

In this paper, we use the following primitive recursive function symbols $x \dot{=} y, \dot{\neg} x, \dot{\exists}xy$, the substitution function $x(y/z)$, and the numeral function $num(x)$ such that PA derives the following:

- $\ulcorner s \urcorner \dot{=} \ulcorner t \urcorner = \ulcorner s = t \urcorner$ for each $\mathcal{L}'$-terms s and t,

Faithful Relative Truth Definability

- $\dot{\neg}\ulcorner\varphi\urcorner = \ulcorner\neg\varphi\urcorner$ for each $\mathcal{L}'$-formula φ,

- $\dot{\exists}\ulcorner v\urcorner\ulcorner\varphi\urcorner = \ulcorner\neg\forall v\neg\varphi\urcorner$ for each $\mathcal{L}'$-variable v and $\mathcal{L}'$-formula φ,

- $\ulcorner\varphi(v)\urcorner(\ulcorner t\urcorner/\ulcorner v\urcorner) = \ulcorner\varphi(t)\urcorner$ for each $\mathcal{L}'$-variable v, $\mathcal{L}'$-formula φ and $\mathcal{L}'$-term t,

- $num(\bar{n}) = \ulcorner\bar{n}\urcorner$ for each $n < \omega$,

where $\ulcorner\varphi(t)\urcorner$ is the result of replacing each occurrence of a free variable v by a term t. The symbols $\dot{\vee}$, $\dot{\wedge}$, and $\dot{\forall}$ are similarly defined. We use an abbreviation $\ulcorner\varphi(\dot{x})\urcorner := (\ulcorner\varphi(v)\urcorner)(num(x)/\ulcorner v\urcorner)$. The evaluation function $\circ$ satisfying $(\ulcorner t\urcorner)^\circ = t$ for a closed $\mathcal{L}'$-term t is not primitive recursive, but it is definable in PA and we introduce this as a function symbol (see, e.g., Feferman, 1991).

Next, we can define several primitive recursive relation symbols such that the following hold in PA:

- $\text{Sent}_{\mathcal{L}'}(x)$ for a Gödel number x of an $\mathcal{L}'$-sentence,

- $\text{Var}(x)$ for a Gödel number x of an $\mathcal{L}'$-variable,

- $\text{Term}(x)$ ($\text{Clterm}(x)$) for a Gödel number x of a (closed) $\mathcal{L}'$-term.

In addition, we use a notation system for ordinal numbers up to ε_0. We have function symbols for ordinal sum $x + y$ and exponentiation 2^x. Moreover, basic relations over the ordinals are definable in PA:

- $\text{On}(x)$ for an ordinal number x,

- $\text{Suc}(x)$ / $\text{Lim}(x)$ for a successor / limit ordinal x,

- $x < y$ for ordinals x and y such that x is less than y.[1]

Using the above notations, we introduce some abbreviations about quantification: v, s and t and α, β, γ and δ range over the codes of variables, closed terms and ordinal numbers, respectively. For example, $\forall v(\sim) :\equiv \forall x(\text{Var}(x) \to (\sim))$, $\forall s(\sim) :\equiv \forall x(\text{Clterm}(x) \to (\sim))$ and $\exists \alpha(\sim) :\equiv \exists x(\text{On}(x) \wedge (\sim))$, and so on. In addition, let $\forall x \in \text{Sent}_{\mathcal{L}'}(\sim) :\equiv \forall x(\text{Sent}_{\mathcal{L}'}(x) \to (\sim))$.

---

[1] For the details of the ordinal notation system, see (Pohlers, 2009). For the representation of syntax in PA I mainly follow (Fujimoto, 2010) and (Halbach, 2014).

Let T be a new unary predicate and $\mathcal{L}_T := \mathcal{L} \cup \{T\}$. In this paper we consider ramified theories of truth, and hence we also introduce the set $\{T_\alpha : \alpha < \varepsilon_0\}$ of truth predicates with indexes and let $\mathcal{L}_{<\alpha} := \mathcal{L} \cup \{T_\beta : \beta < \alpha\}$.

3 Faithful relative truth definability

When can we say that one theory is stronger than or equivalent to the other theory as a theory of truth? Since we need to check consequences containing truth predicates, to have the same $\mathcal{L}$-theorems is necessary, but insufficient as a criteria for the equivalence. For example, the theories $PA + \neg \exists x T(x)$ and $PA + \forall x T(x)$ are obviously conservative over PA, and thus they are arithmetically equivalent. However, They do not seem to have the same concept of truth. In other words, we cannot simply compare them as theories of truth in this case. On the other hand, when a theory S derives all theorems of a theory R, we should be able to say that S has the strength more than or as much as that of R.

In general, the languages of the given theories differ from each other and thus we cannot simply compare their consequences. Furthermore, even if they have the same truth predicates, they may be inconsistent with each other as the above example. However, even in such cases, it is possible that one theory realizes all principles of the other theory via some translation, and then we can regard the former as an extension of the latter. In other words, the former is truth-theoretically stronger than the latter. Fujimoto's relative truth definition captures this insight:

Definition 1 (Relative truth definability) *For $\mathcal{L}_R, \mathcal{L}_S \subseteq \{T\} \cup \mathcal{L}_{<\varepsilon_0}$, let R and S be an $\mathcal{L}_R$- and $\mathcal{L}_S$-theory, respectively. A function $H \colon \mathcal{L}_R \to \mathcal{L}_S$ is a relative truth definition of R into S if H satisfies the following:*

H *replaces each truth predicate T' of $\mathcal{L}_R$ by some $\mathcal{L}_S$-formula θ' with the same free variables, and other symbols are unchanged. Then, for each $\mathcal{L}_R$-formula φ,*

$$R \vdash \varphi \text{ implies } S \vdash H(\varphi).$$

However, relative truth definability itself does not tell us how much the theory is stronger than the other. The theory may be dubious as an adequate truth theory, and thus may be a problematic extension of the other. For example, FS can define the truth predicates of $RT_{<\omega}$, but FS is ω-inconsistent and several authors considers this feature undesirable as an adequate of truth

Faithful Relative Truth Definability

theory. As an extreme case, assume that one theory is inconsistent. Then, this theory trivially defines the truth predicates of the other.

Conversely, when can the above theory S be regarded as weak seen from the theory R? The first idea is as follows: For S and R such that there is a relative truth definition of R into S, S is not so stronger than R if S is $\mathcal{L}$-conservative over R. However, this criteria seems insufficient for the following reason:

Definition 2 *The $\mathcal{L}_T$-theory* TB *consists of* PA *and the set of Tarskian biconditionals:*

$$\varphi \leftrightarrow T(\ulcorner\varphi\urcorner) \quad \text{for each } \mathcal{L}\text{-sentence } \varphi.$$

The $\mathcal{L}_T$-theory TB^+ *is* TB *with the additional axiom:*

$$\neg\forall x \in \text{Sent}_\mathcal{L} \forall y \in \text{Sent}_\mathcal{L}(T(x \veebar y) \leftrightarrow T(x) \vee T(y)).$$

In the same way as Halbach (1999), we can show that TB^+ is conservative over PA, and hence TB^+ is an $\mathcal{L}$-conservative extension of TB. Thus, TB^+ and TB satisfy the above criteria via the identity translation function. On the other hand, since TB can prove $T(\ulcorner\varphi \vee \psi\urcorner) \leftrightarrow T(\ulcorner\varphi\urcorner) \vee T(\ulcorner\psi\urcorner)$ for each $\mathcal{L}$-sentence φ and ψ, TB^+ is ω-inconsistent. Thus TB^+ is problematically stronger than TB, and we need a notion stricter than conservativity.

As one solution, we require that S is a kind of conservative extension of R:

Definition 3 (Faithful Relative Truth Definability) *Let R and S be as in Definition 1, and a function* H *be a relative truth definition of R into S. Then,* H *is a faithful relative truth definition of R into S if for each $\mathcal{L}_R$-formula φ,*

$$R \vdash \varphi \text{ iff } S \vdash H(\varphi).$$

When R and S satisfy this faithfulness by some relative truth definition H, S can be regarded as having the same $\mathcal{L}_R$-consequences as R through H. We construe each theorem of S which is not of the form $H(\varphi)$ for an $\mathcal{L}_R$-sentence φ as a merely formal expression. Based on this understanding, S is truth-theoretically not so stronger than R.

Below, we especially consider faithfulness of ramified theories of truth into type-free theories FS and KF. As a result, we have that the relationship between FS and $RT_{<\omega}$ is not that of TB^+ and TB.

Definition 4 *For $\alpha \leq \varepsilon_0$, the theory $RT_{<\alpha}$ is given by PA with all induction axioms for $\mathcal{L}_{<\alpha}$ and the following axioms for all $\gamma < \beta < \alpha$:*

R1$_\beta$ $\forall s \forall t (T_\beta(s \dot{=} t) \leftrightarrow s^\circ = t^\circ)$,

R2$_\beta$ $\forall x(\text{Sent}_{<\bar\beta}(x) \to (T_\beta(\neg x) \leftrightarrow \neg T_\beta(x)))$,

R3$_\beta$ $\forall x \forall y(\text{Sent}_{<\bar\beta}(x \dot\vee y) \to (T_\beta(x \dot\vee y) \leftrightarrow T_\beta(x) \vee T_\beta(y)))$,

R4$_\beta$ $\forall v \forall x(\text{Sent}_{<\bar\beta}(\dot\forall vx) \to (T_\beta(\dot\forall vx) \leftrightarrow \forall t T_\beta(x(t/v))))$,

R5$_{\beta,\gamma}$ $\forall t(T_\beta(\dot T_{\bar\gamma}(t)) \leftrightarrow T_\gamma(t^\circ))$,

R6$_\beta$ $\forall x(T_\beta(x) \to \text{Sent}_{<\bar\beta}(x))$,

R7$_\beta$ $\forall t \forall \delta < \bar\beta(T_\beta(\dot T_\delta(t)) \leftrightarrow (T_\beta(t^\circ) \wedge \text{Sent}_{<\delta}(t^\circ)))$.[2]

4 Friedman-Sheard theory

The following definition is due to Halbach (1994):

Definition 5 *The theory* FS *is given by* PA *with all induction axioms for* $\mathcal{L}_T$ *and the following axioms and rules:*

FS1 $\forall s \forall t (T(s \dot{=} t) \leftrightarrow s^\circ = t^\circ)$,

FS2 $\forall x(\text{Sent}_{\mathcal{L}_T}(x) \to (T(\neg x) \leftrightarrow \neg T(x)))$,

FS3 $\forall x(\text{Sent}_{\mathcal{L}_T}(x \dot\vee y) \to (T(x \dot\vee y) \leftrightarrow T(x) \vee T(y)))$,

FS4 $\forall v \forall x(\text{Sent}_{\mathcal{L}_T}(\dot\forall vx) \to (T(\dot\forall vx) \leftrightarrow \forall t T(x(t/v))))$,

$$\frac{\varphi}{T(\varphi)} \text{ NEC} \qquad \frac{T(\varphi)}{\varphi} \text{ CONEC} \qquad \textit{for each } \mathcal{L}_T\textit{-sentence } \varphi.$$

In this paper, we use the particular translation $H_{<\alpha}(\varphi)$ such that $H_{<\alpha}(T_\beta(t)) \equiv T(h(t)) \wedge \text{Sent}_{\mathcal{L}_{<\beta}}(t)$ for $\beta < \alpha$ and other symbols are unchanged, where $h(x)$ is a primitive recursive representation of $H_{<\alpha}$.

Fact 1 (Halbach, 1994) *The translation $H_{<\omega}$ is a relative truth definition of* RT$_{<\omega}$ *into* FS, *that is, for each $\mathcal{L}_{<\omega}$-sentence φ:*

---

[2] The definition here is slightly different from (Halbach, 2014). However, this change does not affect the proof-theoretic strength of RT$_{<\alpha}$.

Faithful Relative Truth Definability

$$\mathsf{RT}_{<\omega} \vdash \varphi \text{ implies } \mathsf{FS} \vdash \mathrm{H}_{<\omega}(\varphi).$$

For the converse direction, Halbach defines the translation $\mathrm{G}_n : \mathcal{L}_\mathrm{T} \to \mathcal{L}_{<\omega}$ for each $n < \omega$:

- $\mathrm{G}_0(\mathrm{T}(t)) \equiv 0 = 1$,
- $\mathrm{G}_{n+1}(\mathrm{T}(t)) \equiv \mathrm{T}_n(\mathrm{g}_{\bar{n}}(t))$, where $\mathrm{g}_{\bar{n}}$ is a primitive recursive representation of G_n,
- Other symbols are unchanged.

Then, Halbach shows the following:

Fact 2 (Halbach, 1994) *For each $\mathcal{L}_\mathrm{T}$-sentence φ:*

if $\mathsf{FS} \vdash \varphi$, then $\mathsf{RT}_{<\omega} \vdash \mathrm{G}_n(\varphi)$ for a sufficiently large $n < \omega$.

For the proof of faithfulness, we extend this result as follows.

Lemma 1 *For each $n < \omega$, $\mathcal{L}_{<n}$-formula φ and $l \geq n$,*

$$\mathsf{RT}_{<l} \vdash \mathrm{G}_l(\mathrm{H}_{<\omega}(\varphi)) \leftrightarrow \varphi.$$

Moreover, this is formalizable: For each $n < \omega$ and $l \geq n$,

$$\mathsf{PA} \vdash \forall x \in \mathrm{Sent}_{<\bar{n}}.\mathrm{Bew}_{\mathsf{RT}_{<l}}(\mathrm{g}_{\bar{l}}(\mathrm{h}(x)) \leftrightarrow x),$$

where $\mathrm{Bew}_\mathsf{S}(x)$ is a canonical provability predicate for the theory S.

Proof. We prove by induction on n.

(Basis) Since φ is an $\mathcal{L}$-formula, $\varphi \equiv \mathrm{H}_{<\omega}(\varphi) \equiv \mathrm{G}_l(\mathrm{H}_{<\omega}(\varphi))$, and thus the claim trivially holds.

(Inductive Step) Take any $l \geq n$. We prove by the sub-induction on the complexity of φ. For the base case, let $\varphi \equiv \mathrm{T}_m(t)$ for some $m < n$ and term t. Then $\mathrm{H}_{<\omega}(\varphi) \equiv \mathrm{T}(\mathrm{h}(t)) \wedge \mathrm{Sent}_{\mathcal{L}_{<\bar{m}}}(t)$ and $\mathrm{G}_l(\mathrm{H}_{<\omega}(\varphi)) \equiv \mathrm{T}_{l-1}(\mathrm{g}_{l-1}(\mathrm{h}(t))) \wedge \mathrm{Sent}_{<\bar{m}}(t)$. It is well known that $\mathsf{RT}_{<l}$ proves the global reflection principle for $\mathsf{RT}_{<l-1}$ (see, e.g., Fujimoto, 2010, Proposition 11):

$$\mathsf{RT}_{<l} \vdash \forall x \in \mathrm{Sent}_{<\overline{l-1}}.(\mathrm{Bew}_{\mathsf{RT}_{<l-1}}(x) \to \mathrm{T}_{l-1}(x)).$$

Moreover, we have by main induction hypothesis the following:

$$\mathsf{PA} \vdash \forall x \in \mathrm{Sent}_{<\bar{m}}.\mathrm{Bew}_{\mathsf{RT}_{<l-1}}(\mathrm{g}_{\bar{l}}(\mathrm{h}(x)) \leftrightarrow x).$$

Combining these facts, we obtain:

$$\mathsf{RT}_{<l} \vdash (\mathrm{T}_{l-1}(t) \wedge \mathrm{Sent}_{<\bar{m}}(t)) \leftrightarrow \mathrm{G}_l(\mathrm{H}_{<\omega}(\mathrm{T}_m(t))).$$

Since $l - 1 \geq m$, $\mathsf{RT}_{<l} \vdash (\mathrm{T}_{l-1}(t) \wedge \mathrm{Sent}_{<\bar{m}}(t)) \leftrightarrow \mathrm{T}_m(t)$, and therefore the claim is obtained for $\varphi \equiv \mathrm{T}_m(t)$.

If φ is a complex sentence, then the claim immediately follows from the induction hypothesis. □

From Facts 1 and 2 and Lemma 1 we obtain the following:

Theorem 1 *The translation* $\mathrm{H}_{<\omega}$ *is a faithful relative truth definition of* $\mathsf{RT}_{<\omega}$ *into* FS.

Now that we can see why the case for TB and TB^+ is not applied to $\mathsf{RT}_{<\omega}$ and FS. Assume that $\mathsf{RT}_{<\omega} \vdash \varphi(\bar{n})$ for an $\mathcal{L}_{<\omega}$-formula φ and each $n \in \omega$. Then, $\mathsf{FS} \vdash \mathrm{H}_{<\omega}(\varphi(\bar{n}))$ for each $n \in \omega$. Now if $\mathsf{FS} \vdash \neg\forall x \mathrm{H}_{<\omega}(\varphi(x))$, then since $\neg\forall x \mathrm{H}_{<\omega}(\varphi(x)) \equiv \mathrm{H}_{<\omega}(\neg\forall x \varphi(x))$, we have $\mathsf{RT}_{<\omega} \vdash \neg\forall x \varphi(x)$ by faithfulness of $\mathrm{H}_{<\omega}$. However this contradicts ω-consistency of $\mathsf{RT}_{<\omega}$. Thus $\mathsf{FS} \nvdash \neg\forall x \mathrm{H}_{<\omega}(\varphi(x))$ holds, and hence as far as we consider $\mathrm{H}_{<\omega}$, FS is not a truth-theoretically problematic extension of $\mathsf{RT}_{<\omega}$ in spite of ω-inconsistency of FS.

5 Kripke-Feferman theory

Definition 6 *The theory* KF *is given by* PA *with all induction axioms for* $\mathcal{L}_\mathrm{T}$ *and the following axioms:*

K1 $\forall s \forall t (\mathrm{T}(s \mathbin{\dot{=}} t) \leftrightarrow s^\circ = t^\circ)$,

K2 $\forall s \forall t (\mathrm{T}(\neg s \mathbin{\dot{=}} t) \leftrightarrow \neg s^\circ = t^\circ)$,

K3 $\forall x (\mathrm{Sent}_\mathrm{T}(x) \rightarrow (\mathrm{T}(\dot{\neg}\dot{\neg}x) \leftrightarrow \mathrm{T}(x)))$,

K4 $\forall x \forall y (\mathrm{Sent}_{\mathcal{L}_\mathrm{T}}(x \mathbin{\dot{\vee}} y) \rightarrow (\mathrm{T}(x \mathbin{\dot{\vee}} y) \leftrightarrow \mathrm{T}(x) \vee \mathrm{T}(y)))$,

K5 $\forall x \forall y (\mathrm{Sent}_{\mathcal{L}_\mathrm{T}}(x \mathbin{\dot{\vee}} y) \rightarrow (\mathrm{T}\dot{\neg}(x \mathbin{\dot{\vee}} y) \leftrightarrow \mathrm{T}(\dot{\neg}x) \wedge \mathrm{T}(\dot{\neg}y)))$,

K6 $\forall v \forall x (\mathrm{Sent}_{\mathcal{L}_\mathrm{T}}(\dot{\forall}vx) \rightarrow (\mathrm{T}(\dot{\forall}vx) \leftrightarrow \forall t \mathrm{T}(x(t/v))))$,

K7 $\forall v \forall x (\mathrm{Sent}_{\mathcal{L}_\mathrm{T}}(\dot{\forall}vx) \rightarrow (\mathrm{T}(\dot{\neg}\dot{\forall}vx) \leftrightarrow \exists t \mathrm{T}(x(t/v))))$,

Faithful Relative Truth Definability

K8 $\forall t(T(\underline{T}(t)) \leftrightarrow T(t^\circ))$,

K9 $\forall t(T(\neg \underline{T}(t)) \leftrightarrow T(\neg t^\circ))$.

The $\mathcal{L}_T$-theories KF + CONS and KF + COMP are KF with the following additional axioms, respectively:
CONS $\forall x \in \text{Sent}_{\mathcal{L}_T}(\neg(Tx \wedge T\neg x))$,
COMP $\forall x \in \text{Sent}_{\mathcal{L}_T}(Tx \vee T\neg x)$.

Fact 3 (cf. Halbach, 2014) For each $\mathcal{L}_{<\varepsilon_0}$-sentence φ:

$$\text{RT}_{<\varepsilon_0} \vdash \varphi \text{ implies } \text{KF} \vdash \text{H}_{<\varepsilon_0}(\varphi).$$

Fact 4 (Cantini, 1989) For each $\mathcal{L}$-sentence φ:

$$\text{KF} \vdash \varphi \text{ implies } \text{RT}_{<\varepsilon_0} \vdash \varphi.$$

For the proof of Fact 4 Cantini (1989) gave an infinitary system in which every theorem of KF + CONS is derivable and every cut for complex sentence can be eliminated.[3] Moreover this fact is effectively formalizable in PA, that is, letting $\text{Bew}_\star^\infty(x, y)$ express that (the code of) the sequent y is recursively derivable in the system with the length x and cut-rules only for literal sentences, the following holds:

Fact 5 (Cantini, 1989) For each $\mathcal{L}_T$-sentence φ:

$$\text{KF} \vdash \varphi \text{ implies } \text{PA} \vdash \text{Bew}_\star^\infty(\bar{\beta}, \{\ulcorner\varphi\urcorner\}) \text{ for some } \beta < \varepsilon_0,$$

where $\{x\}$ assigns the Gödel number of the singleton of an $\mathcal{L}_T$-sentence x.

Using primitive recursion theorem we define a 2-ary primitive recursive function symbol $p^y x$ such that $\text{RT}_{<\varepsilon_0}$ proves the following for each $\alpha < \varepsilon_0$:
$\forall \beta < \bar{\alpha} \forall x \forall y \forall s \forall t \forall v.\ ($

$(\beta = 0 \to \neg T_\alpha(p^\beta(x))) \wedge$

$(\text{Suc}(\beta) \to$

- $(T_\alpha(p^\beta(s \doteq t)) \leftrightarrow s^\circ = t^\circ) \wedge$

- $(T_\alpha(p^\beta(\neg(s \doteq t))) \leftrightarrow \neg(s^\circ = t^\circ)) \wedge$

---
[3]For the precise definition of the system, see (Cantini, 1989, p. 123).

- $(T_\alpha(p^\beta(\dot{T}(t))) \leftrightarrow T_\alpha(p^{\beta-1}(t^\circ))) \wedge$

- $(T_\alpha(p^\beta(\neg\dot{T}(t))) \leftrightarrow T_\alpha(p^{\beta-1}(\neg t^\circ)) \vee \neg\mathrm{Sent}_{\mathcal{L}_T}(t^\circ)) \wedge$

- $(\mathrm{Sent}_{\mathcal{L}_T}(x) \rightarrow (T_\alpha(p^\beta(\neg\neg x)) \leftrightarrow T_\alpha(p^{\beta-1}(x)))) \wedge$

- $(\mathrm{Sent}_{\mathcal{L}_T}(x \dot{\vee} y) \rightarrow (T_\alpha(p^\beta(x \dot{\vee} y)) \leftrightarrow (T_\alpha(p^{\beta-1}(x)) \vee T_\alpha(p^{\beta-1}(y)))$
)) $\wedge$

- $(\mathrm{Sent}_{\mathcal{L}_T}(x \dot{\vee} y) \rightarrow (T_\alpha(p^\beta(\neg(x \dot{\vee} y))) \leftrightarrow (T_\alpha(p^{\beta-1}(\neg x)) \wedge T_\alpha(p^{\beta-1}(\neg y))))) \wedge$

- $(\mathrm{Sent}_{\mathcal{L}_T}(\dot{\forall} vx) \rightarrow (T_\alpha(p^\beta(\dot{\forall} vx)) \leftrightarrow \forall t T_\alpha(p^{\beta-1}(x(t/v))))) \wedge$

- $(\mathrm{Sent}_{\mathcal{L}_T}(\dot{\forall} vx) \rightarrow (T_\alpha(p^\beta(\neg\dot{\forall} vx)) \leftrightarrow \exists t T_\alpha(p^{\beta-1}(\neg x(t/v)))))) \wedge$

$(\mathrm{Lim}(\beta) \rightarrow (T_\alpha(p^\beta(x)) \leftrightarrow \exists \delta < \beta. T_\alpha(p^\delta(x)))) \qquad)$.

Again by primitive recursion theorem, we introduce a 3-ary primitive recursive function $h^y_z(x)$ for asymmetric interpretation:

$$h^\alpha_\beta(x) := \begin{cases} h^\alpha_\beta(y) \vee h^\alpha_\beta(z) & (x = y \dot{\vee} z) \\ h^\alpha_\beta(\neg y) \wedge h^\alpha_\beta(\neg z) & (x = \neg(y \dot{\vee} z)) \\ \forall v(h^\alpha_\beta(y)) & (x = \dot{\forall} vy) \\ \exists v(h^\alpha_\beta(\neg y)) & (x = \neg\dot{\forall} vy) \\ x & (x = s \dot{=} t) \\ x & (x = \neg(s \dot{=} t)) \\ h^\alpha_\beta(y) & (x = \neg\neg y) \\ T_{\alpha+1}(\ulcorner p^{\dot\alpha}(v)\urcorner(t/\ulcorner v\urcorner)) & (x = \dot{T}(t)) \\ \neg T_{\beta+1}(\ulcorner p^{\dot\beta}(v)\urcorner(t/\ulcorner v\urcorner)) & (x = \neg\dot{T}(t)) \end{cases}$$

The following lemmata except Lemma 4 are straight formalizations of the proof by Cantini (1989): [4]

---

[4] Note that Nicolai (2018) gives a similar formalization argument in the other theory.

Faithful Relative Truth Definability

Lemma 2 (Formalized Persistency) *Fix any $\alpha < \varepsilon_0$. Then,* $\mathsf{RT}_{<\varepsilon_0}$ *proves:*

$$\forall \delta' < \delta < \beta < \beta' < \bar{\alpha}.\forall x \in \mathsf{Sent}_{\mathcal{L}_T}(\mathsf{T}_{\alpha+1}(\mathsf{h}^\beta_\delta(x)) \to \mathsf{T}_{\alpha+1}(\mathsf{h}^{\beta'}_{\delta'}(x))).$$

Lemma 3 (Formalized Asymmetric Interpretation) *Fix any $\alpha < \varepsilon_0$. Then,*

$$\mathsf{RT}_{<\varepsilon_0} \vdash \forall \delta < \bar{\alpha}.\ \forall y(\mathsf{Bew}^\infty_\star(\delta, y) \to \forall \beta < \bar{\omega}^\alpha.\mathsf{T}_{\omega^\alpha}(\mathsf{h}^{\beta+2^\delta}_\beta(\bigvee y))),$$

where $\bigvee y$ is a primitive recursive function symbol which returns the Gödel number of the disjunction of all sentences in a finite set y.

Proof. The proof is also by a straight formalization of (Cantini, 1989), and thus we omit the detail. However, for the formalization we need to care about the scale of the ordinal numbers used in the proof. Hence, we deal only with the critical case for the interested reader.

The proof is by formal induction on δ up to α. Assume that the last inference of the derivation of y is a cut-rule for a literal sentence $\underline{T}(t)$:

$$\frac{\mathsf{Bew}^\infty_\star(\delta', y \cup \{\underline{T}(t)\}) \qquad \mathsf{Bew}^\infty_\star(\delta', y \cup \{\neg\underline{T}(t)\})}{\mathsf{Bew}^\infty_\star(\delta, y)} \text{cut} \quad \text{for some } \delta' < \delta.$$

Now we have by induction hypothesis the following:

$$\forall \beta < \bar{\omega}^\alpha.\mathsf{T}_{\omega^\alpha}(\mathsf{h}^{\beta+2^{\delta'}}_\beta((\bigvee y) \vee \underline{T}(t))). \tag{1}$$

$$\forall \beta < \bar{\omega}^\alpha.\mathsf{T}_{\omega^\alpha}(\mathsf{h}^{\beta+2^{\delta'}}_\beta((\bigvee y) \vee \neg\underline{T}(t))). \tag{2}$$

We fix any $\beta < \overline{\omega^\alpha}$. Applying Lemma 2 to (1), we have in $\mathsf{RT}_{<\varepsilon_0}$:

$$\mathsf{T}_{\omega^\alpha}(\mathsf{h}^{\beta+2^\delta}_\beta(\bigvee y)) \vee \mathsf{T}_{\omega^\alpha}(\mathsf{h}^{\beta+2^{\delta'}}_\beta(\underline{T}(t))). \tag{3}$$

Next, letting $\beta := \beta + 2^{\delta'} < \overline{\omega^\alpha}$ we obtain from (2) the following:

$$\mathsf{T}_{\omega^\alpha}(\mathsf{h}^{\beta+2^\delta}_{\beta+2^{\delta'}}(\bigvee y)) \vee \mathsf{T}_{\omega^\alpha}(\mathsf{h}^{\beta+2^\delta}_{\beta+2^{\delta'}}(\neg\underline{T}(t))). \tag{4}$$

Again by applying Lemma 2 to (4), we have the following:

$$\mathsf{T}_{\omega^\alpha}(\mathsf{h}^{\beta+2^\delta}_\beta(\bigvee y)) \vee \mathsf{T}_{\omega^\alpha}(\mathsf{h}^{\beta+2^\delta}_{\beta+2^{\delta'}}(\neg\underline{T}(t))). \tag{5}$$

Finally, from (3), (5) the conclusion $T_{\omega^\alpha}(h_\beta^{\beta+2^\delta}(\bigvee y))$ follows. □

Lemma 4 *Fix any* $\alpha < \varepsilon_0$. *Then,*

$$\mathsf{RT}_{<\varepsilon_0} \vdash \forall \beta < \bar{\alpha}. \forall x \in \mathrm{Sent}_{\mathcal{L}_{<\bar{\alpha}}}(T_{\alpha+1}(\mathrm{p}^\beta(\mathrm{h}(x)))) \to T_{\alpha+1}(x)).$$

Proof. We informally work in $\mathsf{RT}_{<\varepsilon_0}$ and implicitly use the properties about $\mathrm{p}^y x$ listed above. The proof is by induction on β.

If $\beta = 0$, then the claim is vacuously true. If β is a limit ordinal, then there exists a $\delta < \beta$ such that $T_{\alpha+1}(\mathrm{p}^\delta(\mathrm{h}(x)))$ holds, and therefore $T_{\alpha+1}(x)$ is true by main induction hypothesis. Finally let β be a successor ordinal. We consider only three cases.

($x = \forall vy$) Then, $\mathrm{h}(x) = \forall v \mathrm{h}(y)$. Therefore, by the property of $\mathrm{p}^y x$, $T_{\alpha+1}(\mathrm{p}^\beta(\mathrm{h}(x)))$ implies $\forall t. T_{\alpha+1}(\mathrm{p}^{\beta-1}(\mathrm{h}(y)(t/v)))$, which is equivalent to $\forall t. T_{\alpha+1}(\mathrm{p}^{\beta-1}(\mathrm{h}(y(t/v))))$. Now we can use sub induction hypothesis and obtain $\forall t. T_{\alpha+1}(y(t/v))$, which is equivalent to $T_{\alpha+1}(x)$.

($x = \underset{\cdot}{T}_\gamma(t)$ for $\gamma < \alpha$) Then $\mathrm{h}(x) = (\ulcorner T(\mathrm{h}(v)) \wedge \mathrm{Sent}_{<\dot{\gamma}}(v)\urcorner)(t/\ulcorner v\urcorner)$, and thus $T_{\alpha+1}(\mathrm{p}^\beta(\mathrm{h}(x))) \leftrightarrow T_{\alpha+1}(\mathrm{p}^{\beta-1}(\underset{\cdot}{T}(\ulcorner \mathrm{h}(v)\urcorner(t/\ulcorner v\urcorner)))) \wedge T_{\alpha+1}(\mathrm{p}^{\beta-1}(\ulcorner \mathrm{Sent}_{<\dot{\gamma}}(v)\urcorner(t/\ulcorner v\urcorner)))$. By induction hypothesis it follows that $\exists \beta' < \beta-1. T_{\alpha+1}(\mathrm{p}^{\beta'}(\mathrm{h}(t^\circ))) \wedge T_{\alpha+1}(\ulcorner \mathrm{Sent}_{<\dot{\gamma}}(v)\urcorner(t/\ulcorner v\urcorner))$. Again by induction hypothesis we obtain $T_{\alpha+1}(t^\circ) \wedge \mathrm{Sent}_{<\gamma}(t^\circ)$, which is equivalent to $T_{\alpha+1}(\underset{\cdot}{T}_\gamma(t))$, as required.

($x = \neg \underset{\cdot}{T}_\gamma(t)$ for $\gamma < \alpha$) If $t^\circ \notin \mathrm{Sent}_{<\gamma}$, then $T_{\alpha+1}(\neg \underset{\cdot}{T}_\gamma(t))$ holds by the axioms of $\mathsf{RT}_{<\varepsilon_0}$. Hence we may assume that $t^\circ \in \mathrm{Sent}_{<\gamma}$, and then by the similar argument as above $T_{\alpha+1}(\mathrm{p}^\beta(\mathrm{h}(\neg \underset{\cdot}{T}_\gamma(t))))$ implies $\exists \beta' < \beta. T_{\alpha+1}(\mathrm{p}^{\beta'}(\mathrm{h}(\neg t^\circ)))$. Therefore by induction hypothesis $T_{\alpha+1}(\neg t^\circ)$ follows. Finally, since $t^\circ \in \mathrm{Sent}_{<\gamma}$ we obtain $T_{\alpha+1}(\neg \underset{\cdot}{T}_\gamma(t))$. □

Theorem 2 *The translation* $\mathrm{H}_{<\varepsilon_0}$ *is a faithful relative truth definition of* $\mathsf{RT}_{<\varepsilon_0}$ *into* $\mathsf{KF} + \mathsf{CONS}$.

Proof. By Fact 3 $\mathrm{H}_{<\varepsilon_0}$ is a relative truth definition of $\mathsf{RT}_{<\varepsilon_0}$ into $\mathsf{KF} + \mathsf{CONS}$, and hence we show the faithfulness. Assume that $\mathsf{KF} \vdash \mathrm{H}_{<\varepsilon_0}(\varphi)$ for $\alpha < \varepsilon_0$ and an $\mathcal{L}_{<\alpha}$-sentence φ. Since $\mathsf{RT}_{<\varepsilon_0} \vdash T_\alpha(\ulcorner \varphi \urcorner) \leftrightarrow \varphi$ and $\mathrm{H}_{<\varepsilon_0}$ is a relative truth definition of $\mathsf{RT}_{<\varepsilon_0}$ into KF, KF also proves $T(\mathrm{h}(\ulcorner \varphi \urcorner))$.

Faithful Relative Truth Definability

Then by Fact 5 we have $\text{RT}_{<\varepsilon_0} \vdash \text{Bew}_*^\infty(\bar\beta, \{\underline{T}(\ulcorner h(\ulcorner\varphi\urcorner)\urcorner)\})$ for some $\alpha < \beta < \varepsilon_0$. Next, by formalized aymmetric interpretation we obtain $\text{RT}_{<\varepsilon_0} \vdash T_{\omega^{\beta+1}}(h_0^{\overline{2^\beta}}(\underline{T}(\ulcorner h(\ulcorner\varphi\urcorner)\urcorner)))$. Thus $\text{RT}_{<\varepsilon_0} \vdash T_{\omega^{\beta+1}}(T_{\overline{2^\beta+1}}(\ulcorner p^{\overline{2^\beta}}(v)\urcorner(\ulcorner h(\ulcorner\varphi\urcorner)\urcorner/\ulcorner v\urcorner)))$ follows by the definition of $h_z^y(x)$. Since $2^\beta + 1 < \omega^{\beta+1}$, we also have $\text{RT}_{<\varepsilon_0} \vdash T_{\omega^{\beta+1}}(p^{\overline{2^\beta}}(h(\ulcorner\varphi\urcorner)))$. Thus, we obtain by Lemma 4 $\text{RT}_{<\varepsilon_0} \vdash T_{\omega^{\beta+1}}(\ulcorner\varphi\urcorner)$, and hence $\text{RT}_{<\varepsilon_0} \vdash \varphi$. □

Fact 6 (Cantini, 1989) *Let a translation* $G : \mathcal{L}_T \to \mathcal{L}_T$ *replace each occurrence of* $T(t)$ *by* $\neg T(\neg t)$. *Then,*

$$\text{KF} + \text{COMP} \vdash \varphi \text{ implies } \text{KF} + \text{CONS} \vdash G(\varphi).$$

Theorem 3 *The translation* $H_{<\varepsilon_0}$ *is also a faithful relative truth definition of* $\text{RT}_{<\varepsilon_0}$ *into* $\text{KF} + \text{COMP}$.

Proof. We show that $\text{KF} + \text{CONS} \vdash G(H_{<\varepsilon_0}(\varphi)) \leftrightarrow H_{<\varepsilon_0}(\varphi)$ for each $\mathcal{L}_{<\varepsilon_0}$-formula φ by induction on the complexity of φ. Then, the claim immediately follows from Theorem 2 and Fact 6.

We consider the case where $\varphi \equiv T_\alpha(t)$ for $\alpha < \varepsilon_0$ and a term t. Now $H_{<\varepsilon_0}(\varphi) \equiv T(h(t)) \land \text{Sent}_{<\alpha}(t)$ and $G(H_{<\varepsilon_0}(\varphi)) \equiv \neg T(\neg h(t)) \land \text{Sent}_{<\alpha}(t)$. Since $\text{KF} + \text{CONS} \vdash H_{<\varepsilon_0}(R2\alpha)$, that is, $\text{KF} + \text{CONS} \vdash \forall x(\text{Sent}_{<\alpha}(x) \to (T(h(\neg x)) \leftrightarrow \neg T(h(x))))$, we have $\text{KF} + \text{CONS} \vdash \text{Sent}_{<\alpha}(t) \to (\neg T(h(\neg t)) \leftrightarrow T(h(t)))$. As a result, we obtain $\text{KF} + \text{CONS} \vdash G(H_{<\varepsilon_0}(T_\alpha(t))) \leftrightarrow H_{<\varepsilon_0}(T_\alpha(t))$.

When φ is a complex formula, the claim holds by induction hypothesis. □

6 Concluding remarks

By faithful relative truth definability of R into S, S can be seen as an instrument for deriving the theorems of R. For example, theorems of $\text{FS}_{<\omega}$ which are not in the range of the given faithful relative truth definition are regarded as merely formal expressions, and then $\text{FS}_{<\omega}$ is justified as a theory of truth once $\text{RT}_{<\omega}$ is accepted. In this way other type-free theories of truth may also be justified in terms of other theories.

Note that Fujimoto's motivation for relative truth definability is also related to a criteria of conceptual equivalence between two truth theories.

Roughly speaking, the proposal of Fujimoto (2010) is that two truth theories have conceptualy equivalent truth when they are mutually relatively truth definable. This condition is not satisfied for $RT_{<\omega}$ and FS, nor $RT_{<\varepsilon_0}$ and KF. This means that conceptual equivalence as mutual relative truth definability is not necessary for truth-theoretic justification via faithfulness.

Conversely, does (should) conceptual equivalence imply justification via faithfulness? One trivial example is given when two theories are identical. Obviously, any theory is conceptually equivalent (in any natural sense) to and faithfully relative truth definable into itself. Similarly, if we take conceptual equivalence as merely expressional difference like "wahr" and "true" rather than some equivalence of consequences like conservativity, then conceptual equivalence seems to imply faithfulness. Nicolai (2017) recently raises a doubt about mutual relative truth definability as a criteria of conceptual equivalence and gives a stricter criteria of conceptual equivalence, τ-equivalence. Roughly speaking, τ-equivalence demands that the composition of the two mutual relative truth definitions is (provably) isomorphic to the identity function, and then the one theory functions as a faithful mirror for the other theory, and vice versa (Nicolai, 2017, pp. 337–338). Therefore, if this faithfulness relation is required for conceptual equivalence, then Nicolai's criteria may seem to be more adequate than Fujimoto's.

Finally, we note the possibility of faithfulness in other theories. The first is an extension KF^* of KF, the schematic reflective closure (Feferman, 1991). This is known to be reducible to $RT_{<\Gamma_0}$, and thus we can expect faithfulness of $RT_{<\Gamma_0}$ into KF^*. The second is faithfulness for other arithmetical theories. Nicolai (2017) extends the notion of τ-equivalence to e-equivalence in order to include subsystems of second order arithmetic, and then he proves faithfulness-like results of $RT_{<n}$ into $RA_{<n}$. Therefore, from the justificational point of view faithfulness results may make progress Halbach's program on ontological reduction of set existence assumptions into semantic assumptions (Halbach, 2000). The final example is Feferman's determinate theory of truth DT (Feferman, 2008). Since DT as with KF has a problem that the inner and the outer logic of the truth predicate do not coincide, Feferman suggests that DT should be understood instrumentally and for this purpose one may need more than consistency or conservativity over $RA_{<\varepsilon_0}$ (Feferman, 2008, p. 215). Thus, if we can prove faithfulness of $RT_{<\varepsilon_0}$ or $RA_{<\varepsilon_0}$ into DT, then we would get a more desirable relative justification of DT.

References

Cantini, A. (1989). Notes on formal theories of truth. *Zeitschrift für mathematische Logik und Grundlagen der Mathematik, 35*(2), 97–130.

Feferman, S. (1991). Reflecting on incompleteness. *The Journal of Symbolic Logic, 56*(1), 1–49.

Feferman, S. (2008). Axioms for determinateness and truth. *The Review of Symbolic Logic, 1*(2), 204–217.

Friedman, H., & Sheard, M. (1987). An axiomatic approach to self-referential truth. *Annals of Pure and Applied Logic, 33*, 1–21.

Fujimoto, K. (2010). Relative truth definability of axiomatic truth theories. *Bulletin of Symbolic Logic, 16*(3), 305–344.

Halbach, V. (1994). A system of complete and consistent truth. *Notre Dame Journal of Formal Logic, 35*(3), 311–327.

Halbach, V. (1999). Disquotationalism and infinite conjunctions. *Mind, 108*(429), 1–22.

Halbach, V. (2000). Truth and reduction. *Erkenntnis, 53*(1-2), 97–126.

Halbach, V. (2014). *Axiomatic Theories of Truth*. Cambridge University Press.

Nicolai, C. (2017). Equivalences for truth predicates. *The Review of Symbolic Logic, 10*(2), 322–356.

Nicolai, C. (2018). Provably true sentences across axiomatizations of Kripke's theory of truth. *Studia Logica, 106*(1), 101–130.

Pohlers, W. (2009). *Proof Theory: The First Step into Impredicativity*. Berlin, Heidelberg: Springer.

Daichi Hayashi
Hokkaido University, Faculty of Letters
Japan
E-mail: `dhayashi@eis.hokudai.ac.jp`

Toward a Logic for Neural Networks

LEVIN HORNISCHER[1]

Abstract: Neural networks and related computing systems suffer from the notorious *black box problem*: despite their success, we lack a general framework or language to reason about the behavior of these systems. We need a logic with a mathematical semantics for this. In this paper, we sketch a first such logic: a mathematical structure with a logic that describes the behavior of possibly non-deterministic, discrete dynamical systems (which include neural networks). The mathematical structure is based on *domain theory*. Domains solved the 'black box problem' of (classical) computers by providing a denotational semantics for computer programs. Can it analogously be used for neural networks? We show that, under precise conditions, the possible behaviors of a system form a domain—relating domain theoretic concepts to properties of the system. This mathematical structure can interpret the well-behaved logic HYPE which thus can be used to reason about both the long-term behavior and the history of the system.

Keywords: neural networks, dynamical systems, black box problem, domain theory, logic, HYPE, unbounded nondeterminism

1 Introduction

Among computing systems, neural networks gained great prominence in the recent years due to their success in image recognition, natural language processing, and, more generally, in learning from great amounts of data. However, despite their success, they suffer from the notorious *black box problem*: we don't fully understand how they do what they do. Making neural networks and modern artificial intelligence more transparent has been dubbed 'explainable AI'. While explainable AI recently made progress in understanding *specific* applications of neural networks, we still lack a general framework or language to reason about the behavior of these systems. That is, we need a logic together with a mathematical semantics that is able

---

[1]For inspiring discussions, I'm grateful to Samson Abramsky, Franz Berto, Jon Michael Dunn, Michiel van Lambalgen, Hannes Leitgeb, and the audience at Logica 2018.

to describe the behavior of these computing systems. In this paper, we want to sketch a first such logic.

In section 2, we define the class of computing systems that we'll consider. In section 3, we define the set of possible behaviors of such a system and a partial order on this set. In section 4, we present precise conditions on the system ensuring the behavior poset to be an algebraic domain in the sense of domain theory. In section 5, we then show how this domain can be transformed in a structure that interprets the recently developed logic HYPE. We conclude in section 6.

The aim of the paper is to convey the main idea of this 'logic plus semantics' for neural nets. Thus, we (necessarily have to) focus on the intuition behind it, leaving the details for elsewhere (Hornischer, 2018). We indicate possible further development of this idea and potentially fruitful connections to other fields (e.g., automata theory, coalgebra, homology, or general relativity).

2 Dynamical systems and trajectories

We define the notion of a possibly non-deterministic discrete dynamical system that we'll be working with, and we show that both neural network computation and learning are such systems.

In the most general sense, a *possibly non-deterministic discrete dynamical system* is a pair (S, f) where S is a non-empty set, called the *state space*, and $f : S \to \mathcal{P}(S) \setminus \{\emptyset\}$ is a function mapping each state s to a non-empty set of states, called the *successor states* of s. Intuitively, S is the set of states that the system can be in, and f is the local rule describing which states the system might evolve to given the current state. (Thus, these systems can be seen as *transition systems* known in computer science.) If there always is exactly one successor state, the system is called *deterministic*.[2] A (finite or infinite) *trajectory* is a (finite or infinite) sequence $\langle s_0, s_1, \ldots \rangle$ of states in S such that for each $i \geq 0$, s_{i+1} is a successor state of s_i. We denote the empty trajectory by $\bot$. Intuitively, a trajectory is a possible evolution of the system (a path through the state space). For brevity we call a possibly non-deterministic discrete dynamical systems a *(non-) deterministic system*.

This definition is very general. Depending on the kind of application,

---

[2] If we want to be even more general, we could allow the local update rule to change over time, that is, define a system as $(S, \{f_n\}_{n \in \mathbb{N}})$ where each f_n maps states to non-empty sets of states. However, for convenience we stick to the simpler definition.

Toward a Logic for Neural Networks

dynamical systems are usually assumed to have additional structure: that the state space carries a topology and the dynamics is continuous or that the state carries a measure and the dynamics is measure preserving. However, we need to work with the general definition if we want to capture all the different kinds of neural networks and related computing systems. It is surprising that, as we'll see, even though this general notion of a dynamical system has very little structure, a lot of structure will emerge when considering equivalence classes of trajectories.

Let's see that computation in neural networks can be regarded as (non-) deterministic system. Roughly, a *neural network* is a collection of neurons that are linked via synapses. Each neuron can have an activation that describes whether and how the neuron is firing. Each synapse connecting neuron i to neuron j carries a weight describing how much activation can pass from i to j. A state of the network describes the activation of all the neurons at a given time. The activation propagates through the synapses and determines the successor state: if neurons $i_1, \ldots, i_k$ are connected by synapses to neuron j, then the activation of j at time $n+1$ is computed from the activation of $i_1, \ldots, i_k$ at time n. Thus, a neural network is a deterministic system (S, f) where S is the set of all possible states of activation of the neurons and f describes the propagation of activation. Of course, much more can be said, e.g., about the different kinds of neural nets (feed forward, recurrent, etc.) or about learning (changing of weights).[3] But, again to remain as general as needed, we don't need further details for now. In fact, many (if not most) computing systems can be seen as (non-) deterministic systems (Turing machines, automata, etc.).

On this perspective, neural networks are deterministic systems—so why did we include non-deterministic systems? For two reasons. First, since usually continuous (rather than binary) activation values are allowed, the state space is infinite (of size continuum). Thus, to understand the behavior of the neural net, it is useful to partition the state space into finitely many cells: e.g., clustering observationally equivalent states together. The dynamics induced on these cells may be non-deterministic: a cell can contain two states whose successor states are in two different cells. Second, instead of understanding the computation of a trained neural network, we may also wish to describe the training of a neural network. That is, instead of sequences of neuron-activation given by the propagation rule we consider sequences

---
[3] Details can be found in the many textbooks on neural networks. Albeit older, Rojas (1996) has a good presentation of the mathematical basics.

of network-weights given by a learning algorithm—and learning might be non-deterministic.

3 Behavioral equivalence and order

If we want to analyze the behavior of systems, we should have a notion of what a possible behavior of a system is. A trajectory of a system is an instance of a possible behavior, but two trajectories might exhibit the same behavior. Thus, a possible behavior is a trajectory modulo behavioral equivalence. So we need to define when two trajectories t and t' are behaviorally equivalent. Intuitively, t and t' should agree on the computationally important information. That is, in the case of neural networks,

(i) t and t' yield the same result: there is a (minimal) point where the two trajectories meet and continue the same path.

(ii) t and t' gather the same information along the way: they described the same cycles and visited the same stable states before reaching the meeting point.

The second point is motivated by the rule of thumb from dynamical systems theory that in the stable states and cycles of a system lies the information computed by the system. For example, a Hopfield neural network retrieves memorized pictures in its stable states, and the limit cycle in a Hodgkin-Huxley model of a neuron describes the spiking pattern of the neuron.

Let's formalize this intuitive idea of behavioral equivalence. Concerning (i), call a pair $(i, j) \in \mathbb{N}$ a *locally minimal coincidence pair* of two trajectories t and t' if $t(i) = t'(j)$ and, whenever defined,

$$t(i-1) \neq t'(j-1) \qquad t(i-1) \neq t'(j) \qquad t(i) \neq t'(j-1).^4$$

Say t and t' have the *same tail* starting in (i, j) (denoted $t(i) \ldots = t'(j) \ldots$), if for all n, $t(i+n)$ is defined iff $t'(j+n)$ is defined and in both cases they are equal. Then (i) says that there is a locally minimal coincidence pair (i, j) starting in which t and t' have the same tail.

---

[4]For a globally minimal coincidence pair we don't quantify only about the direct precursors but over all precursors. This doesn't make sense for non-deterministic systems, and for deterministic systems one can show that the local and the global formulation of the resulting notion of behavioral equivalence are equivalent.

Toward a Logic for Neural Networks

Concerning (ii), we need to additionally formalize that $t(0) \ldots t(i)$ and $t'(0) \ldots t'(j)$ are *cycle equivalent* and *stable state equivalent*. We do this first explicitly and then axiomatically.

If s' is a successor state of s, call the transition from s to s' a *cycle edge* if there is a trajectory from s' back to s. Say two finite trajectories t and t' are *cycle equivalent* if the cycle edges occurring on t are precisely the cycle edges occurring on t', and the respective number of occurrences are the same, too. The *occurrence profile* of a stable state s (i.e., s is a successor of itself) on a finite trajectory t is a finite (possibly empty) sequence of non-negative integers $\langle n_1, \ldots, n_k \rangle$ where there are k-many blocks of uninterrupted repetitions of s on t such that $n_i \geq 1$ is the number of repetitions of s in the i-th block.[5] Say two finite trajectories t and t' are *stable state equivalent* if the stable states occurring on t are precisely the stable states occurring on t', and the stable state have the same occurrence profile on t and t', respectively.

In fact, we'll rely only on a few of the properties of these definitions. Thus, for transparency and to allow our results to be applicable to a range of possible explicit definitions, we introduce cycle- and stable state equivalence axiomatically. We call equivalence relations $\approx_c$ and $\approx_s$ on finite trajectories *cycle equivalence* and *stable state equivalence*, respectively, if they satisfy the following axioms. (For readability we omit the qualifier 'and the trajectory is possible in the system'.)

1. Both $ta \approx_c t'a$ iff $tat'' \approx_c t'at''$, and $ta \approx_s t'a$ iff $tat'' \approx_s t'at''$.

2. If $a \neq b$, then $taa \not\approx_s t'ba$, $ta \not\approx_s taat'$, and $ta \not\approx_c tabt'a$.

3. If $aa_1 \ldots a_n a$ occurs on t but no a_i on t', then $t \not\approx_c t'$.

4. If there is no stable state in t and t', then $ta \approx_s t'a$.

5. If there is no limit cycles in system, then $\approx_c$ is trivial.

6. If $t \approx_c t'$, then for every occurrence of $aa_1 \ldots a_n a$ on t, there is an occurrence of some a_i on t'.

7. If $t \approx_s t'$, then for every occurrence of stable state a on t, there is an occurrence of a on t'.

---
[5]For example, the occurrence profile of stable state a on trajectory $aabab$ is $\langle 2, 1 \rangle$.

The axioms are satisfied by the brute force definition. More elegant definitions might be found by considering homology on (the simplex graph of) the system considered as graph.[6]

For the remainder of the paper, we fix two relations $\approx_c$ and $\approx_s$ satisfying the above axioms. We summarize.

Definition 1 (Behavioral equivalence) *Given a (non-) deterministic system, two trajectories t and t' are behaviorally equivalent ($t \equiv t'$) if $t = \bot = t'$ or there is a locally minimal coincidence pair (i, j) of (t, t') such that $t(i) \ldots = t'(j) \ldots$, $t(0) \ldots t(i) \approx_c t'(0) \ldots t'(j)$, and $t(0) \ldots t(i) \approx_s t'(0) \ldots t'(j)$. We write $\mathbb{T} := \{[t] : t \text{ is a trajectory}\}$ for the set of equivalence classes of trajectories.*

Thus, $\mathbb{T}$ is the set of possible behaviors of the given (non-) deterministic system. We next observe that there is a natural order on $\mathbb{T}$: $[t] \leq [t']$ if any trajectory equivalent to t can be extended to one equivalent to t', that is,

$$[t] \leq [t'] \text{ iff } \forall t_0 \in [t] \, \exists t_1 \in [t'] : t \preceq t'$$

(iff, intuitively, behavior $[t]$ can be extended by the system into behavior $[t']$). It follows from the axioms on $\approx_c$ and $\approx_s$ that $(\mathbb{T}, \leq)$ is a partial order (it trivially is a preorder).

In the next sections, we'll investigate the order of behaviors $(\mathbb{T}, \leq)$.

4 Long-term behavior and history as a domain

We'll provide precise conditions on the system such that the order of behaviors $(\mathbb{T}, \leq)$, capturing the long-term behavior of the system, and its dual, capturing the history of the system, are algebraic domains.

We start with a very brief recap of domain-theory. Domain theory was originated by Dana Scott and others in the late 1960's (Scott, 1970). Since then, it developed into a rich logico-mathematical framework to understand computation (Abramsky & Jung, 1994). Roughly, domain theory studies particular partial orders that can be regarded as "information orderings": the elements represent pieces of information or partially known objects. As one moves up in the order, one gains more information (about the objects). The important formal concepts are the following. Let $(P, \leq)$ be a partial order.

---

[6]Generally speaking, a homological perspective on neural networks is fruitful (see, e.g., Reimann et al., 2017).

Toward a Logic for Neural Networks

A subset $D \subseteq P$ is *directed* if $D \neq \emptyset$ and any two elements of D have an upper bound in D. Moreover, P is a *directed-complete* partial order (dcpo) if every directed subset of P has a least upper bound. Thus, intuitively, chains of ever increasing information converge to a limit. If P is a dcpo, we say that x is *way below* y ($x \ll y$) if if any directed set whose least upper bound is above y already contains an element above x. An element $x \in P$ is *compact* if $x \ll x$, so x is, in a sense, finitary. An *algebraic domain* is a dcpo where, for every element x, the compact elements below x form a directed set whose least upper bound is x. Algebraic domains are particularly well behaved since their important properties are already determined by the compact elements.

Most importantly, domain theory developed denotational semantics for computer programs (of various programming languages). So it provides mathematical objects explaining what a program is computing. Can it also be used to analogously analyze the behavior of computing systems? Yes, as we'll now show: for appropriate systems, the set of behaviors is an algebraic domain.

Theorem 1 (Upward trajectory domain) *Let $(\mathbb{T}, \leq)$ be the set of behaviors of a (non-) deterministic system. $(\mathbb{T}, \leq)$ is an algebraic domain if the state space doesn't contain the subsystems in Figure 1 (arrows indicate paths and not necessarily direct successors).*[7] *We then refer to $(\mathbb{T}, \leq)$ as* upward trajectory domain. *(The converse holds under some more assumptions on $\approx_c$ and $\approx_s$.)*

Intuitively, what the forbidden subsystems have in common is that there are infinitely many choices: In the left-most one, there are infinitely many ways to go from a_1 to b. In the middle one, there are infinitely many ways of changing from the a-path to the b-path. In the right-most system, there are infinitely many choices between doing the b-cycle or the c-cycle. Thus,

---

[7]The following side conditions are imposed: In the left-most subsystem, (i) all indicated states are distinct. (ii) There is no state that is both reachable from all a_n's and reaches some a_n. And (iii) there is an infinite trajectory t through all a_n's and a finite trajectory t' ending in b such that $[t']$ is an upper bound of $[t \restriction 1] \leq [t \restriction 2] \leq \dots$.

In the middle subsystem, (i) none of the a_n's is identical to another one, but arbitrarily many b_n's might be identical and some b_n's might be equal to some a_n. (ii) If t is a trajectory through all a_n's, then $[t \restriction 1] \leq [t \restriction 2] \leq \dots$ doesn't have a finite upper bound. And (iii) there is an infinite trajectory t through all a_n's and an infinite trajectory t' through all b_n's such that $[t']$ is an upper bound of $[t \restriction 1] \leq [t \restriction 2] \leq \dots$ and $t' \not\equiv t$.

In the right-most subsystem, either $a \neq b$ or $a \neq c$ (or both).

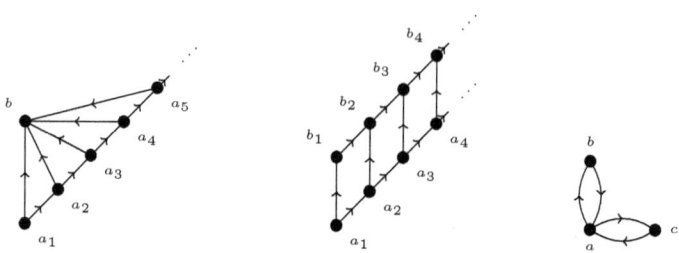

Figure 1: Forbidden subsystems (upward)

if this 'infinite non-determinism'[8] is excluded, the set of behaviors is an algebraic domain.

The upward trajectory domain describes the long-term behavior of the system: the further one moves up in the order starting at a finite $[t]$, the further the continuation of behavior of $[t]$ one considers. Since it is a domain, it can then be analyzed with the tools of domain theory (Abramsky & Jung, 1994). For example, all finite behaviors are compact, whence algebraicity entails that the global (i.e., infinite) behavior of the system can be completely described from observing finite behavior.[9] Moreover, domain theoretic concepts can be related to properties of the system. The theorem indicates such a connection between algebraicity and 'finite nondeterminism.' One could conjecture other connections: $(\mathbb{T}, \leq)$ is a Scott domain or a bifinite domain if additionally the left-most and, respectively, the middle subsystem of Figure 2 is excluded.

While the upward trajectory domain $(\mathbb{T}, \leq)$ describes the structure of the long-term behavior of the system – that is, the *effects* of certain states –, we're also often interested in the *causes* of a state. Then we need to look at the history of the system: the order-dual $\leq^{\mathrm{op}}$ of $\leq$. We can again find conditions on the system guaranteeing this partial order to be an algebraic domain, too.

Theorem 2 (Downward trajectory domain) *For a (non-) deterministic sys-*

---

[8]This is not the exactly the same as the well-known term 'unbounded non-determinism' which usually means that some state contains infinitely many successor states (as, e.g., in the diamond of Figure 2).

[9]Formally, an algebraic domain is isomorphic to the ideal completion of its compact elements.

Toward a Logic for Neural Networks

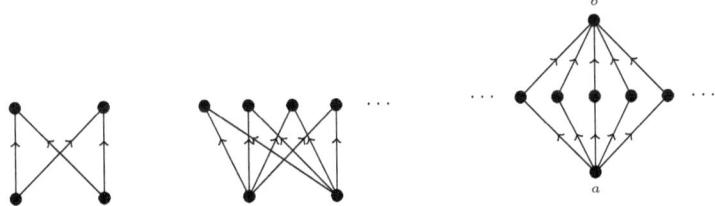

Figure 2: Forbidden subsystems related to bounded completeness, bifiniteness, and interval-compactness.

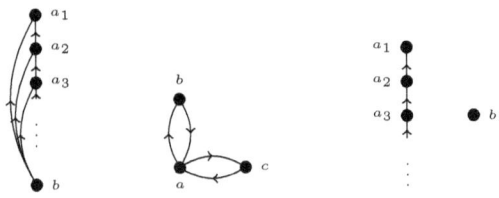

Figure 3: Forbidden subsystem (downward)

tem, $(\mathbb{T}, \leq^{op})$ *is an algebraic domain if the state space doesn't contain the subsystems in Figure 3 (again, arrows indicate paths and not necessarily direct successors).*[10] *We then refer to* $(\mathbb{T}, \leq)$ as downward trajectory domain.

We'll next consider how to combine the upward and downward order into a new structure that can interpret the logic HYPE.

5   The combined trajectory domain and a logic

As mentioned, when we investigate a system, we're both interested in the limit-behavior (effects) and the history (causes). Thus, in particular, when we reason about the limit-behavior of the system, we thus should also take into account from where the limit was reached. This suggests to replace

---

[10]The following side conditions are imposed: In the right-most subsystem, there are trajectories $t', t_1, t_2, \ldots$ ending in $b, a_1, a_2, \ldots$, respectively, such that $[t_1] > [t_2] > \ldots$ and, for all n, $[t_n] \not\leq [t']$.

(infinite) limit-behaviors by new elements representing the different possibilities from where these limit-behaviors can be reached. We achieve this by essentially mirroring $\mathbb{T}$ at the infinite trajectories. To be precise, let t be a trajectory. Then the set

$$\llbracket t \rrbracket := \{ [t'] : t' \text{ infinite trajectory and } t \preceq t' \}$$

records the possible limit behavior of t. If t is infinite, $\llbracket t \rrbracket = \{[t]\}$, whence we identify it with $[t]$. To get the refined limit states that keep track of the origin from where they were reached, we add the new elements $(\llbracket t \rrbracket, [t])$ to $\mathbb{T}$ for all finite $[t]$. Qua refined limits, the new elements are above the respective old ones: If $[t'] \in \mathbb{T}$, then $[t'] \leq (\llbracket t \rrbracket, [t])$ iff $[t']$ and $[t]$ share a common limit-behavior, that is, there is $[t_m] \in \llbracket t \rrbracket$ such that $[t'] \leq [t_m]$. The new elements are ordered among each other as follows: the further away the origin from which the limit (represented by the new element) was reached, the further up is the new element. That is, $(\llbracket t \rrbracket, [t]) \leq (\llbracket t' \rrbracket, [t'])$ iff $[t'] \leq [t]$. For reasons of symmetry, we can also think of the old elements $[t]$ as really being of the form $([t], \llbracket t \rrbracket)$. We summarize.

Definition 2 (Combined trajectory domain) *For a (non-) deterministic system, define $\mathbb{T}_c$ as the union of $\{\bot, \top\}$ and*

$$L := \{([t], \llbracket t \rrbracket) : \bot \neq t \text{ finite trajectory}\},$$
$$U := \{(\llbracket t \rrbracket, [t]) : \bot \neq t \text{ finite trajectory}\},$$
$$M := \{(\llbracket t \rrbracket, \llbracket t \rrbracket) : t \text{ infinite trajectory}\},$$

called the lower half, upper half, *and the* limit trajectories *respectively. Define the order $\leq_c$ on $\mathbb{T}_c$ as follows ($\leq$ is the order on $\mathbb{T}$):*

$$([t], \llbracket t \rrbracket) \leq_c ([t'], \llbracket t' \rrbracket) \quad \text{if } [t] \leq [t']$$
$$(\llbracket t \rrbracket, [t]) \leq_c (\llbracket t' \rrbracket, [t']) \quad \text{if } [t] \geq [t']$$
$$([t], \llbracket t \rrbracket) \leq_c (\llbracket t' \rrbracket, [t']) \quad \text{if } \exists [t_m] \in \llbracket t' \rrbracket : [t] \leq [t_m],$$

and $\bot$ and $\top$ are the $\leq_c$-least and biggest element, respectively. The order for limit trajectories $[t]$ is given via the above by the identification $\llbracket t \rrbracket = [t]$. We call $(\mathbb{T}_c, \leq_c)$ the combined trajectory domain *of the system.*

An involution poset $(P, \leq, ')$ is a poset $(P, \leq)$ with a function $\cdot' : P \to P$ such that $x'' = x$ and $x \leq y$ implies $y' \leq x'$. We observe that it is an involution to switch perspectives from regarding a trajectory as finite trajectory to regarding some of its limit behavior as being reached from that trajectory.

Toward a Logic for Neural Networks

Proposition 1 *Let $(\mathbb{T}_c, \leq_c)$ be the combined trajectory domain of a (non-)deterministic system. Define a mapping $\cdot^* : \mathbb{T}_c \to \mathbb{T}_c$ by*

$$((t, [\![t]\!]))^* := ([\![t]\!], t) \quad \text{and} \quad (([\![t]\!], t))^* := (t, [\![t]\!]).$$

Then $\cdot^*$ is an involution. Moreover, if τ is finite, $\tau \leq_c \tau^*$, and if τ is infinite, $\tau^* = \tau$.

We now can get to the logic. Leitgeb (2018) recently developed a general logical system called HYPE. Although not originally intended, we show that this logic can, in fact, be interpreted by dynamical systems.[11]

We first recap HYPE (Leitgeb, 2018). Fix a propositional language with variables $p_1, p_2, \ldots$ and connectives $\neg, \wedge, \vee, \to$. Let Lit be the set of literals (negated or non-negated propositional variables). Given a literal l, its dual is denoted $\bar{l}$.[12] A valuation on a nonempty set M is a function $V : M \to \mathcal{P}(\text{Lit})$. A HYPE *model* $\mathfrak{M}$ (for our fixed propositional language) is a quadrupel $(M, V, \circ, \perp)$ where M is a nonempty set, V a valuation on M, and the following axioms hold—for the detailed formulation of the axioms see (Leitgeb, 2018):

- $\circ$ is a partial binary function from $M \times M$ to M (the *fusion function*) such that V is $\circ$-monotone and $\circ$ is reflexive, commutative, and weakly associative.

- $\perp$ is a binary symmetric relation on M (the *incompatibility relation*) such that literal incompatibility entails $\perp$, and $s \perp s'$ entails $s \circ s'' \perp s'$, whenever defined.

- For every $s \in M$ there is a unique $s^* \in M$ (the *star image of s*) such that $V(s^*) = \{\bar{v} : v \notin V(s)\}$, $s^{**} = s$, $s \not\perp s^*$, and s^* is the $\circ$-largest state compatible with s.

We define $s \leq s'$ if $s \circ s'$ exists and $s' = s \circ s'$. Formula satisfaction and logical consequence in HYPE models are defined as follows (Leitgeb, 2018):

- $s \models v$ iff $l \in V(s)$ (where l is a literal).

- $s \models \neg \varphi$ iff for all s', if $s' \models \varphi$, then $s \perp s'$.

---

[11] A hint that such an interpretation is possible is the HYPE model consisting of fixed-points of the untyped truth-predicate (Leitgeb, 2018)—and fixed-point operators are a general form of computation.

[12] So if $l = p$, then $\bar{l} = \neg p$, and if $l = \neg p$, then $\bar{l} = p$.

(A1)	$\vdash \top$	(A10)	$\vdash (A \to C) \to ((B \to C)$
(A2)	$\vdash A \to A$		$\to (A \vee B \to C))$
(A3)	$\vdash A \to (B \to A)$	(A11)	$\vdash A \wedge (B \vee C) \leftrightarrow (A \wedge B) \vee (A \wedge C)$
(A4)	$\vdash A \to (B \to C) \to$	(A12)	$\vdash A \vee (B \wedge C) \leftrightarrow (A \vee B) \wedge (A \vee C)$
	$((A \to B) \to (A \to C))$	(A13)	$\vdash A \leftrightarrow \neg\neg A$
(A5)	$\vdash A \wedge B \to A$	(A14)	$\vdash \neg(A \wedge B) \leftrightarrow \neg A \vee \neg B$
(A6)	$\vdash A \wedge B \to B$	(A15)	$\vdash \neg(A \vee B) \leftrightarrow \neg A \wedge \neg B$
(A7)	$\vdash A \to A \vee B$	(A16)	$\dfrac{\vdash A \to B}{\vdash \neg B \to \neg A}$
(A8)	$\vdash B \to A \vee B$		
(A9)	$\vdash A \to (B \to A \wedge B)$	(A17)	$A, A \to B \vdash B$

Figure 4: The system HYPE as presented by Leitgeb (2018).

- $s \models \varphi \wedge \psi$ and $s \models \varphi \vee \psi$ as usual.

- $s \models \varphi \to \psi$ iff for all s', if $s' \models \varphi$ and $s \circ s'$ is defined, then $s \circ s' \models \psi$.

Given a set Γ of formulas and a formula ψ, $\Gamma \models \psi$ iff for all HYPE models $\mathfrak{M}$ and states s of $\mathfrak{M}$, if $s \models \varphi$ for all $\varphi \in \Gamma$, then $s \models \psi$.

Leitgeb (2018) provides a sound and complete logic for HYPE models with this semantics. We'll refer to this logic as HYPE which we repeat for convenience in Figure 4. The logic HYPE is well-understood and well-behaved. It not only is sound and complete (via a canonical model construction) with respect to the HYPE models, it also has the deduction theorem, the disjunction property, the finite model property, and is decidable. Moreover, it contains first-degree entailment, conservatively extends intuitionistic logic, and the structures of HYPE models are well-known from ordinary mathematics. (For all these results see Leitgeb, 2018.)

To show that the combined trajectory domain carries the structure of a HYPE model, we'll make use of the fact that it has an involution (proposition 1).

Proposition 2 (Involution posets as HYPE models) *Let $(M, \leq, ^*)$ be a nonempty involution poset. Let $V : M \to \mathcal{P}(\mathrm{Lit})$ be a valuation such that*

(i) *V is monotone ($s \leq s'$ implies $V(s) \subseteq V(s')$),*

(ii) *If $l \in V(s)$ and $\bar{l} \in V(s')$, then $s \not\leq s'^*$, and*

(iii) *$V(s^*) = \{\bar{l} : l \notin V(s)\}$.*

Toward a Logic for Neural Networks

Then $(M, V, \circ, \perp)$ is a HYPE model where $s \circ s'$ is the least $\leq$-upper bound of $\{s, s'\}$, if it exits, and $s \perp s'$ iff $s \not\leq s'^*$ (so $\perp$ is orthogonality in the involution poset).

So it remains to find appropriate valuations on the combined trajectory domain $(\mathbb{T}_c, \leq_c)$. We do so by lifting a valuation on the state space to a valuation of trajectories as follows.

We start with a valuation $W : S \to \mathcal{P}(\mathsf{Lit})$ of the state space of the system. This captures the idea that we can measure the system with respect to the properties $p_1, \neg p_1, p_2, \neg p_2, \ldots$. That is, any state of the system carries the information of which properties certainly obtain and which certainly do not obtain at that current state of the system. So $[\![l]\!]_W := \{s : l \in W(s)\}$ describes an area of the state space where l holds (or is made true).[13]

We say that W is *separated* if for all p, there is no state that is reachable from some state in $[\![p]\!]_W$ and from some state in $[\![\neg p]\!]_W$.[14] We say that W is *bifurcating* if for all p and trajectories t, if t is undecided on p, then both p and $\neg p$ can be forced. That is, if no trajectory equivalent to t contains a p- or $\neg p$-state, then t can be extended to trajectories t_0 and t_1 such that t_0 is equivalent to a trajectory containing a p-state and t_1 is equivalent to a trajectory containing a $\neg p$-state.

How do we get from a valuation W of the state space to the valuation V of trajectories? The most straightforward idea is to take the valuation of a (lower half) trajectory as collecting all the properties of the states that it has visited. For a limit behavior we collect all the properties that can occur in that limit-behavior.

$$V_W([t], [\![t]\!]) := \bigcup_{t_0 \in [t]} \bigcup_{n \in \mathbb{N}} W(t_0(n))$$

$$V_W([\![t]\!], [t]) := \bigcup_{[t_0] \in [\![t]\!]} \bigcup_{n \in \mathbb{N}} W(t_0(n)).$$

We call the valuation $V_W : \mathbb{T}_c \to \mathcal{P}(\mathsf{Lit})$ defined in this way the *valuation induced by W*. We now have the promised result.

Theorem 3 (Logic of $\mathbb{T}_c$) *Let $(\mathbb{T}_c, \leq_c)$ be the combined trajectory domain of a (non-) deterministic system. Let $W : S \to \mathcal{P}(\mathsf{Lit})$ be a separated and*

---

[13] If the state space S carries a topology, it is natural to demand $[\![l]\!]_W$ to be open—this captures the idea that l is verifiable by finitary measurements (Vickers, 1989).

[14] In other words, if from a state s it is possible that the system develops into a state that is also reached from a $\neg p$ state, then p doesn't "certainly" obtain at s. Rather, p cannot be conclusively decided at state s.

bifurcating valuation of the state space. Then $(\mathbb{T}_c, V_W, \circ, \bot)$ *is a* HYPE *model where* V_W *the valuation induced by* W, $\tau \circ \tau' = \tau \vee \tau'$, *if it exits, and* $\tau \bot \tau' \Leftrightarrow \tau \not\leq \tau'^*$.

What do the logical operations mean? For example, a finite trajectory makes $\neg \varphi$ true iff φ is false in the limit (as reached from this trajectory). And $\varphi \to \psi$ is true at a finite trajectory $[t]$ if any future state that is reachable through a φ-area of the state space is also reachable through a ψ-area. (This is equivalent to any infinite $t_m \in [\![t]\!]$ making the classical conditional $\neg \varphi \vee \psi$ true.) An interesting open question is whether HYPE is also complete for this trajectory domain interpretation (and, if not, which extension of HYPE is).

6 Conclusion and further developments

We started with the aim of developing a logic together with a mathematical semantics to reason about both the long-term behavior and the history of a system. We provided a first such instance with the trajectory domain as mathematical structure and HYPE as logic.

We end with three areas of further development. First, since the general notion of a (non-) deterministic system that we've worked with is a transition system, it's natural to exploit their coalgebraic character. In addition to the homological considerations mentioned above, this might also provide more structural definitions of behavioral equivalence (bisimulation). More generally, it seems worth further exploring the link between the two classical subjects of computer science – transition systems and domain theory – that our results provide.

Second, in the spirit of explainable AI, we might wish to show that for every computing system there is a 'transparent' system with the same behavior. In our framework, this would translate into the following—so to speak the *Hauptvermutung* of the project. For every appropriate (non-) deterministic system S, there is another 'transparent' system S' whose trajectory domain is isomorphic (or otherwise closely related) to that of S. (Note the analogy to DFA minimization in automata theory.)

Third, the order structures found in the trajectory domains are, surprisingly, similar to order structures found in the causality order of spacetimes (in general relativity). This can be seen as follows. Define $\mathbb{T}_{\text{fin}}$ as the finite and nonempty elements of $\mathbb{T}$ end exclude the subsystems of Theorems 1 and 2, and exclude the 'infinite diamond' of Figure 2. Then $(\mathbb{T}_{\text{fin}}, \leq)$ is a causal set in the sense of the causal set approach to quantum gravity (Bombelli, Lee, Meyer, & Sorkin, 1987). Do the excluded subsystems have physical meaning under this interpretation? Moreover, $(\mathbb{T}_{\mathsf{fin}}, \leq)$ then also is a globally hyperbolic poset in the sense of Martin and Panangaden (2006) and thus in the same category as the causality order of globally hyperbolic spacetimes. Does this relate cosmic censorship (valid on globally hyperbolic spacetimes) to bounded non-determinism (valid when diamond excluded)?

References

Abramsky, S., & Jung, A. (1994). Domain theory. In S. Abramsky, D. M. Gabbay, & T. S. E. Maibaum (Eds.), *Handbook of Logic in Computer Science*. Oxford: Clarendon Press.

Bombelli, L., Lee, J., Meyer, D., & Sorkin, R. D. (1987). Space-time as a causal set. *Physical Review Letters, 59*(5), 521–524.

Hornischer, L. (2018). *Trajectory domains: describing the behavior of computing systems*. (Unpublished manuscript.)

Leitgeb, H. (2018). Hype: A system of hyperintensional logic (with an application to semantic paradoxes). *Journal of Philosophical Logic*.

Martin, K., & Panangaden, P. (2006). A domain of space-time intervals in general relativity. *Communications in Mathematical Physics, 267*, 563–586.

Reimann, M. W., Nolte, M., Scolamiero, M., Turner, K., Perin, R., Chindemi, G., ... Markram, H. (2017). Cliques of neurons bound into cavities provide a missing link between structure and function. *Frontiers in Computational Neuroscience, 11*, 48.

Rojas, R. (1996). *Neural Networks: A Systematic Introduction*. Berlin: Springer.

Scott, D. (1970). *Outline of a Mathematical Theory of Computation* (Tech. Rep. No. PRG02). Oxford University Computing Laboratory.

Vickers, S. (1989). *Topology via Logic*. Cambridge: Cambridge University Press.

Levin Hornischer
University of Amsterdam
Institute for Logic, Language and Computation
The Netherlands
E-mail: `l.a.hornischer@uva.nl`

The Name of the Sinus Function

ANSTEN KLEV[1]

Abstract: What is the name of the sinus function? Mathematical symbolism offers two obvious alternatives: $\sin(x)$, including the (independent) variable x; or sin, where no x occurs. It will be argued here that a good mathematical symbolism should give room for both alternatives and, moreover, that each alternative corresponds to a distinct notion of function.

Keywords: functions, mathematical notation, type theory, Gottlob Frege

1 Introduction

Two items that are likely to feature on any list of the central notions in mathematics are the notion of a set, or domain, of objects and the notion of a function from one set to another. The notion of property is of course also central, but following Frege and Russell, properties are usually regarded as functions of a special kind, viz. propositional functions (which in classical logic may be identified with functions into truth-values). Since both items would seem to be equally central to mathematics, it may be surprising to find that philosophers of mathematics rarely pay much attention to the notion of function, whereas the notion of set has been, and continues to be, a subject of intense philosophical research.

Perhaps one feels that the notion of function is in no need of special scrutiny, since it can be reduced to the notion of set: in set theory a function f from A to B is usually defined as a set of ordered pairs $\langle a, b \rangle \in A \times B$ such that for all $a \in A$ there is a unique $b \in B$ with $\langle a, b \rangle \in f$. Ordered pairs themselves can be treated as sets in well-known ways. But although this account of the notion of function may suffice for the technical purposes of set theory, it is inadequate as a philosophical explication. A set in the sense of standard axiomatic set theory is an individual, and as such it is not fit for

---

[1] I have learned about the different notions of function discussed in this paper from the writings and lectures of Per Martin-Löf and Göran Sundholm (see, e.g., Sundholm, 2012, pp. 953–4). Comments from Joan Bertran on a distant predecessor of the paper helped me in the preparation of the current version. Financial support was offered by grant nr. 17-18344Y from the Czech Science Foundation, GAČR.

being applied to an argument. It is in the nature of being a function f from A to B that it can be applied to any $a \in A$ as argument so as to yield an object $f(a) \in B$ as result; but it does not make sense to speak of the application of, say, a number or a book or a stone or any other individual to an argument in this way. The category error thus involved in calling a set a function is revealed by the need in the language of set theory to invoke an application function for the formation of the term $f(a)$. The syntax of predicate logic requires that the term $f(a)$ be subject to function/argument analysis, so we must regard it as the result of applying a certain binary function $\_\__1(\_\__2)$ to f as first argument and a as second argument. (If one wants to reserve the parentheses for proper function application, one may, following Shoenfield, 1967, p. 245, write the term in question as $f'a$, thus employing the apostrophe as a binary function symbol.) It is easy to see by a regress argument similar to Bradley's regress that this application function cannot in turn be treated as a set, whence it seems that set theory has not, after all, managed to get rid of the notion of function.

We are thus led to treat functions as *sui generis*, and so the task arises of throwing light on this kind of mathematical entity. I aim to do so here via a consideration of functional notation. In particular, I wish to consider the two styles of functional notation exemplified by the following two names of the sinus function:

$$\sin(x) \qquad (1)$$

either with parentheses, as here, or without, as in $\sin x$; and

$$\sin \qquad (2)$$

where no variable occurs. I will argue, firstly, that a good mathematical symbolism needs both styles of notation; and, secondly, that each style of notation corresponds to a distinct notion of function. The difference between these two notions of function will be clarified by means of type theory. The paper ends with some remarks on Frege's functional notation.

2 Historical remarks

Style (1) notation, that is, the style of notation exemplified by '$\sin(x)$', is the traditional one. It can be found, for instance, in the famous definition of the notion of function given by Euler:

> A function of a variable quantity is an analytical expression composed in any way from this variable quantity and numbers

Name of the Sinus Function

> or constant quantities. Therefore, every analytical expression, in which apart from the variable quantity z all quantities composing the expression are constant, will be a function of z. Thus $a + 3z$; $az - 4zz$; $az + b\sqrt{aa - zz}$; c^z; etc. are functions of z. (Euler, 1748, p. 18)

Euler was not alone in calling a function an expression: also well-known definitions by Lagrange, Boole, and Frege (in the *Begriffsschrift*) classify functions as expressions.[2] These mathematicians clearly cannot have meant by an expression what a formalist might mean by it, a certain formal object devoid of sense, say. Rather, their idea must have been that a function is a certain law of correspondence given by the meaning of the expression; or that the expression—$f(x)$, say—stands for an unspecified quantity determined by this correspondence, a quantity that gets fully specified once the "variable quantity" x has been specified. In any event, the resulting notation is one in which variables feature essentially; for a variable is needed in order to indicate that a function $f(x)$ either is or determines a dependent quantity, namely one depending on x.

Style (2) notation, that is, the style of notation exemplified by 'sin', is the more modern one. Characteristic of this notation is that the function name occurs without an accompanying free variable: the function is treated as a self-standing object. A good example of this style of notation can be found in Dedekind's definition of what he called a mapping (*Abbildung*):

> By a mapping φ of a system S one understands a law according to which for every determinate element s of S there belongs a determinate thing, which is called the image of s and denoted by $\varphi(s)$; we also say that $\varphi(s)$ corresponds to the element s, that $\varphi(s)$ results or is produced from s by the mapping φ, or that φ transforms s into $\varphi(s)$. (Dedekind, 1888, nr. 26)

Dedekind's choice of the term 'mapping' (also used by Riemann) suggests that he did not wish to think of a function as a dependent quantity. Rather, a function f was, for him, something that transforms an object a into an object $f(a)$, or—as one often says today—that sends the object a to the object $f(a)$. If variables are present in the language (as they will be in most dialects of mathematical language), we get a function in style (1) notation from a function f in style (2) notation by applying f to x, yielding $f(x)$.

---

[2] See the definitions of the notion of function compiled by Rüthing (1984).

3 The need for both styles of notation

Both styles of functional notation are needed in a sufficiently rich mathematical symbolism. One can get a long way with style (1) notation, but once higher-order functions are admitted—functions whose arguments are themselves functions—style (2) notation is almost forced upon one.[3] On the other hand, style (1) notation is very convenient when defining functions.

Let us try to make sense of higher-order functions in style (1) notation. In the first instance, then, we wish to find a name for a second-order function along the lines of $f(x)$. Since we are restricted to style (1) notation, any argument to a second-order function, being a first-order function, has to be written as $f(x)$. The name that then perhaps first suggests itself is $F(f(x))$, since this treats $f(x)$ as argument just as $f(x)$ itself treats x as argument. This cannot, however, be the name of a second-order function, but must rather be regarded as the name of a composite first-order function: the function $F(f(x))$ has been obtained from the first-order function $F(y)$ by a "change of variables", viz. by substituting $f(x)$ for y. For instance, if $f(x)$ is $x^2 + 1$, and $F(y)$ is $\sin(y)$, then $F(f(x))$ is $\sin(x^2 + 1)$.

The definite integral from a to b is a second-order function whose argument is usually written in style (1) notation:

$$\int_a^b f(x)\,dx$$

This cannot be mistaken for a composite first-order function, since here the variable x has become bound, as indicated by the dx following the argument $f(x)$. The variable x must, indeed, somehow be done away with if we are to get an individual as the result of applying a second-order function to $f(x)$. Taking a hint from Frege (1893, § 25), we are therefore led to the notation

$$F_x(f(x))$$

for a second-order function in style (1) notation. This portrays the second-order function as a variable-binding operation: the subscripted x indicates that it has become bound in $F_x(f(x))$.

This may be a satisfactory notation for second-order functions; but it is not clear how to extend it to higher orders. What seems to be the most straightforward generalization to third order is

$$\mathscr{F}_f(F_x(f(x)))$$

---

[3] This point has previously been made by Curry and Feys (1958, ch. 3B).

Name of the Sinus Function

The subscripted f here is meant to indicate that it has become bound in $\mathscr{F}_f(F_x(f(x)))$. But should this subscript really be f, which seems to be style (2) notation; should it not rather be $f(x)$? If that were the subscript, however, it would appear that a third-order function applied to $F_x(f(x))$ as argument binds the whole $f(x)$, and then we could no longer see x as being bound already in $F_x(f(x))$. Neither option seems satisfactory.

One reason why it is difficult to make good sense of the application of a third-order function in style (1) notation is that the argument to such a function as portrayed in this notation is not a pure function, but a function and a variable-binder at once. If we could separate out the variable-binding part of such an impure function, higher-order functions would be easy to handle, since then the application of a higher-order function would be no different from the application of a first-order function: both would be instances of a general notion of function application. As is well-known, we can separate out variable-binding by employing λ-abstraction, writing the result of binding the variable x in $f(x)$ as $\lambda x. f(x)$.[4] This is, however, a function written in style (2) notation. Indeed, by an extensionality principle known in the λ-calculus as rule η, the function $\lambda x. f(x)$ is definitionally identical to f. The purification of style (1) notation by means of λ thus leads us to style (2) notation, where the application of higher-order functions indeed poses no special problems.

We should not be led by these considerations to abandon style (1) notation altogether. The need for style (1) notation can be seen by considering the ordinary practice of defining functions. Suppose I wish to define a function that squares a number and adds 5; how would I give the definition in standard mathematical symbolism? Clearly, I would say something along the lines of: let f by defined by

$$f(x) := x^2 + 5$$

Likewise, to define the boolean function of exclusive or I would say: let $\oplus$ be defined by

$$p \oplus q := (p \vee q) \wedge \neg(p \wedge q)$$

In each case I introduce the function I wish to define in style (2) notation, but the defining equation is written using style (1) notation (this way of proceeding can be seen to rely on the rule η). It is true that by the use of

---

[4] In order to emphasize that abstraction is not itself a function, a notation such as $[x].f(x)$ might be preferable to $\lambda x.f(x)$.

so-called combinators we can get rid of variables altogether;[5] but the most natural and most convenient method of defining functions, viz. the method just illustrated, does employ variables.

4 Different notions of function

It seems clear from the wording of Euler's and Dedekind's definitions that they understood by a function quite different things. Euler's conception is that of a variable quantity depending on another variable quantity, the dependence being given by an "analytical expression". Dedekind's conception is that of a transformation of an object into, or a mapping of an object onto, an object (in the typical case: *another* object). We are thus led to ask whether the two styles of functional notation illustrated by these definitions perhaps correspond to different notions of function. That we indeed have two notions of function here, and not merely two styles of notation, can be seen by the help of type theory.

The difference is apparent already in the simple hierarchy of types. Recall that in this hierarchy we have a certain number of ground types and, moreover, higher types formed by the rule: if α and β are types, then so is $(\alpha)\beta$, viz. the type of functions from α to β. An object f of type $(\alpha)\beta$ is subject to the typing rule: if a is of type α, then $f(a)$, the result of applying f to a, is of type β. Here application is a primitive notion native to types of the form $(\alpha)\beta$. It is clear that a function as portrayed by style (2) notation is an object of type $(\alpha)\beta$. In particular, if we regard the sinus function as defined on the real numbers, $\mathbb{R}$, style (2) notation portrays it as an object of type $(\mathbb{R})\mathbb{R}$. Using the colon to indicate type membership we can write

$$\sin : (\mathbb{R})\mathbb{R}$$

We shall call this form of statement a type assignment or, following Martin-Löf (1984), a judgement.[6]

As already noted, a function in style (1) notation arises from one in style (2) notation by application to a variable: f applied to x yields $f(x)$.

---

[5]The invention of combinators and the idea of using them to get rid of variables goes back to Schönfinkel (1924); that one can define λ-abstraction by means of combinators is shown in (Curry & Feys, 1958, ch. 6).

[6]In (Klev, 2018) the term 'type predication' is used. There it is emphasized that these statements are not of function/argument form, hence they fall outside the frame of ordinary predicate logic.

Name of the Sinus Function

In simple type theory a variable ranging over type α is itself deemed to be of type α. For instance, a variable ranging over $\mathbb{R}$ is itself of type $\mathbb{R}$. For a variable x ranging over the real numbers, the above typing rule therefore yields the judgement

$$\sin(x) : \mathbb{R}$$

It is then obvious that the functions sin and $\sin(x)$ must be quite different things, since they belong to different types: the first is of type $(\mathbb{R})\mathbb{R}$, the second of type $\mathbb{R}$

It is somewhat counterintuitive that $\sin(x)$ is assigned the type $\mathbb{R}$, which is not a type of functions, but of real numbers; it is, for instance, the type assigned as well to $\sin(0)$ and π. A more satisfactory type-theoretical treatment of functions in style (1) notation can be achieved by invoking the notion of a context. By a context is here understood a finite sequence of type assignments to variables:

$$x_1 : \alpha_1, \ldots, x_n : \alpha_n$$

In the classical formulation of type theory by Church (1940) variables are typed globally: a variable has the same type wherever it appears. In a type theory with contexts, by contrast, variables are typed only locally: a judgement $a : \alpha$ must always be provided with a context Γ in which each variable free in a (or α) is assigned a type. When occurring in other judgements, these variables may be assigned other types. We write[7]

$$a : \alpha \dashv \Gamma$$

This may be read as "a is of type α in the context Γ". That there is a difference between $\sin(0)$ and $\sin(x)$ is now clear from the fact that different forms of judgement are required for expressing their type:

$$\sin(0) : \mathbb{R}$$
$$\sin(x) : \mathbb{R} \dashv (x : \mathbb{R})$$

In the first judgement no context is needed, since no variables are present. In the second judgement, by contrast, the context $x : \mathbb{R}$ must be provided. Whereas the first judgement is read simply as "$\sin(0)$ is a real number", the second is read as "$\sin(x)$ is a real number in the context $x : \mathbb{R}$", which

---

[7] The standard notation is $\Gamma \vdash a : \alpha$, but we shall prefer the inversion of this for reasons that will soon become clear.

in turn may be glossed as "sin(x) is a real number provided/whenever/on condition that x is a real number".

Employing traditional logico-grammatical terminology we may say that in a type assignment the colon is the copula; what is to the left of the colon is the subject; and what is to the right of the colon is the predicate. The difference between sin(0), sin(x), and sin may then be expressed by saying that a different predicate occurs in the type assignment of each of these. Exploiting the fact that one of the meanings of *katēgoria* in Aristotle's Greek is predicate, we may say that sin(0), sin(x), and sin are objects of different categories: the category of sin(0) is $\mathbb{R}$, that of sin(x) is $\mathbb{R} \dashv (x : \mathbb{R})$, and that of sin is $(\mathbb{R})\mathbb{R}$.[8]

We have thus found a system of classification that distinguishes the function sin(x) from the function sin and that, moreover, distinguishes these from real numbers such as sin(0) and π. The difference between the two categories $(\alpha)\beta$ and $\beta \dashv (x : \alpha)$ may be brought out further by considering the notion of application appropriate to functions of each category. We cannot call something a function unless we are able to make sense of the application of such a thing to an argument. In set theory the application of a function *qua* set of ordered pairs is accounted for by the invocation of an application function. We saw that for function types $(\alpha)\beta$ application is primitive. The application of a function of category $\beta \dashv (x : \alpha)$, by contrast, is explained in terms of a prior notion of value assignment to variables. Thus we apply $f(x)$ to an argument a by substituting a for x. Using standard substitution notation, the result of applying sin(x) to π, say, can therefore be written as

$$\sin(x)[x := \pi]$$

By means of the rules of substitution, this can be seen to be definitionally identical to sin(π). The latter is of course also the result of applying sin to π, but in that case no intermediate steps of substitution are required.

5 Frege

In Frege's ideography (Frege, 1893), the argument places with which a function name is associated always have to be filled, be it by a closed or an

---

[8]This use of the term 'category' has been suggested by Martin-Löf in lectures. It does, of course, not coincide with the use of the term in the field of mathematics known as category theory.

Name of the Sinus Function

open term. When Frege employs function names in his informal expository language he fills the argument places by the Greek letters ξ and ζ, thus using these letters as argument place indicators. In Frege's ideography we shall therefore find variants of 'sin(x)' and 'sin(π)', and in his expository language variants of 'sin(ξ)'; but nowhere shall we find just 'sin' alone. Indeed, an isolated function name, as Frege calls it, is an *Unding*, a monstrosity (Frege, 1903, p.148): the name 'sin', for instance, purports to be a function name, but without an expression filling or indicating the associated argument place, it appears to be the name of an individual, a proper name, in Frege's terminology.

Frege may thus be said to have been a staunch defender of style (1) notation and an equally staunch critic of style (2) notation. Frege's understanding of style (1) notation was, to be sure, very different from Euler's (cf. Frege, 1904). He did not think of a function as a variable, or indeterminate, quantity nor as a dependent object. Instead he saw the essence of the function in what he called its unsaturatedness or incompleteness. But he held that this unsaturatedness must be visible in the notation and that an isolated function name such as 'sin' fails to make it visible. "The isolation of the function sign contradicts the nature of the function, which consists in its unsaturatedness" Frege wrote in a letter to Russell (Frege, 1976, p. 243, letter dated 13.11.1904). The argument places associated with a function name must always be recognizable as such, either by the presence of argument expressions, as in the ideography, or by the presence of argument place indicators, as in the expository language. The result is a style (1) notation.

Frege's insistence on style (1) notation is nicely illustrated a bit later in this letter to Russell, where he introduces a notational device that in effect amounts to λ-abstraction. He seeks a normal form for function names where the argument place indicator has been placed at the end. The name 'sin ξ' is already in this form, but '$\xi^2 + 5$' is not. Frege employs the *spiritus asper* and writes[9]

$$\grave{\varepsilon}(\varepsilon^2 + 5)\xi$$

This notation is to be understood so that '$\grave{\varepsilon}(\varepsilon^2 + 5)a$' has the same meaning as '$a^2 + 5$' for any suitable argument a. The name '$\grave{\varepsilon}(\varepsilon^2 + 5)$' is thus just what we would write by means of the λ-notation as $\lambda x. x^2 + 5$. Accepting

---

[9] Note that $\grave{\varepsilon}(\varepsilon^2 + 5)$ is not the course-of-values of the function $\xi^2 + 5$, which Frege writes, using the *spiritus lenis*, as '$\acute{\varepsilon}(\varepsilon^2 + 5)$'. Frege employed the *spiritus asper* also in an earlier letter to Russell, but then with an entirely different meaning (Frege, 1976, p. 218, letter dated 29.6.1902). The similarity to λ-abstraction of the *spiritus asper* as used by Frege in the 1904-letter was noted by Nomoto (2006, p. 94); see also Martin-Löf (2006).

function names such as these would therefore mean adopting style (2) notation. Such a notation would be quite appropriate to Frege's ideography, which is a higher-order language. Frege insisted, however, that '$\acute{\varepsilon}(\varepsilon^2 + 5)$' cannot be a function name, since no expression fills or indicates the associated argument place as required by his style (1) notation.

References

Church, A. (1940). A formulation of the simple theory of types. *The Journal of Symbolic Logic*, 5(2), 56–68.

Curry, H. B., & Feys, R. (1958). *Combinatory Logic*. Volume 1. Amsterdam: North-Holland.

Dedekind, R. (1888). *Was sind und was sollen die Zahlen?* Braunschweig: Vieweg und Sohn.

Euler, L. (1748). *Introductio in analysin infinitorum*. Lausanne: Bousquet.

Frege, G. (1893). *Grundgesetze der Arithmetik I*. Jena: Hermann Pohle.

Frege, G. (1903). *Grundgesetze der Arithmetik II*. Jena: Hermann Pohle.

Frege, G. (1904). Was ist eine Funktion? In *Festschrift Ludwig Boltzmann. Gewidmet zum sechzigsten Geburtstage* (pp. 656–666). Leipzig: Johann Ambrosius Barth.

Frege, G. (1976). *Wissenschaftlicher Briefwechsel*. Hamburg: Felix Meiner Verlag.

Klev, A. (2018). The concept horse is a concept. *The Review of Symbolic Logic*, 11, 547–572.

Martin-Löf, P. (1984). *Intuitionistic Type Theory*. Naples: Bibliopolis.

Martin-Löf, P. (2006). Comments on Prof. Kazuyuki Nomoto's paper. *Annals of the Japan Association for Philosophy of Science*, 14, 98–99.

Nomoto, K. (2006). The methodology and structure of Gottlob Frege's logico-philosophical investigations. *Annals of the Japan Association for Philosophy of Science*, 14, 73–97.

Rüthing, D. (1984). Some definitions of the concept of function from Joh. Bernoulli to N. Bourbaki. *Mathematical Intelligencer*, 6, 72–77.

Schönfinkel, M. (1924). Bausteine der mathematischen Logik. *Mathematische Annalen*, 92, 305–316.

Shoenfield, J. R. (1967). *Mathematical Logic*. Reading, MA: Addison-Wesley.

Sundholm, B. G. (2012). "Inference versus consequence" revisited: inference, consequence, conditional, implication. *Synthese, 187*, 943–956.

Ansten Klev
Czech Academy of Sciences, Institute of Philosophy
The Czech Republic
E-mail: `klev@flu.cas.cz`

Algebraic Semantics for Fischer Servi's Version of Modal Logic **BK**

SERGEI ODINTSOV[1]

Abstract: Fischer Servi style logic $\mathbf{BK}^{FS}$ was defined by Odintsov and Wansing (2017) via a standard translation of modal formulas into Belnap-Dunn fist order logic. In this paper, $\mathbf{BK}^{FS}$ will be characterized in terms of twist structures over bimodal algebras. It turs out that the abstract closure of the family of twist structures forms a variety providing an equivalent algebraic semantics for the global consequence relation of $\mathbf{BK}^{FS}$.

Keywords: twist structure, equivalent algebraic semantics, lattice of logics, FDE-based modal logic

1 Introduction

Unlike other approaches to defining intuitionistic modal logics, which are usually based on relational semantics combining in a suitable way accessibility relations used to interpret intuitionistic implication and modal operators, Fischer Servi (1984) defined her intuitionistic modal logic FS syntactically, with the help of the standard translation ST_x from modal language to the language of first order logic. A modal formula φ belongs to FS iff its standard translation is a theorem of first order intuitionistic logic. An unexpected consequence of this definition is the complexity of FS. The logic FS has finite model property (it was established by Grefe, 1998), but the complexity of a counter model for φ is estimated by Ackerman function from the length of φ (see Simpson, 1994). Analysing different approaches to modal logics over FDE (First Degree Entailment) Odintsov and Wansing (2017) introduced Fischer Servi's version $\mathbf{BK}^{FS}$ of the FDE-based modal logic BK with classical weak implication $\rightarrow$, which was suggested in (Odintsov & Wansing, 2010). The definition of $\mathbf{BK}^{FS}$ was based on the standard translation from the language of BK to the language of Belnap-Dunn first order logic (see Sano & Omori, 2014). The logic $\mathbf{BK}^{FS}$ was axiomatized in

---

[1]The author acknowledges the support of the Russian Foundation for Basic Research, project No. 18-501-12019-DFG-a.

(Odintsov & Wansing, 2017) and characterized via a suitable class of birelational models. It is typical that semantics of logics with strong negation connective $\sim$ in the language (like FDE) involves two forcing relations (for verification and for falsification of formulas) and that only strong equivalence $\Leftrightarrow$ defined via the weak equivalence as $\varphi \Leftrightarrow \psi := (\varphi \leftrightarrow \psi) \wedge (\sim\varphi \leftrightarrow \sim\psi)$[2] has congruence property, the congruence property of weak implication $\leftrightarrow$ w.r.t. strong negation $\sim$ usually fails. An unusual feature of Fischer Servi style logic BK^{FS} is that verification and falsification of formulas starting with $\Diamond$-modality are defined via different accessibility relations, as a consequence $\Box$ and $\Diamond$ are not interdefinable via strong negation and strong equivalence. In this paper we investigate the question: whether BK^{FS} is algebraizable. To this end we develop a twist-structure semantics for BK^{FS} (a semantics of this kind for BK was suggested in Odintsov & Wansing, 2010). Then we prove that the abstract closure of the family of twist structures characterizing BK^{FS} forms a variety. Finally, we prove that this variety provides an equivalent algebraic semantics for BK^{FS}.

2 The logic BK^{FS}, relational semantics and Hilbert style calculi

We will work with the propositional modal language

$$\mathcal{L} = \{\vee, \wedge, \rightarrow, \bot, \sim, \Box, \Diamond\}$$

with symbol $\sim$ for strong negation, $\bot$ for falsity constant, and $\rightarrow$ for classical weak implication. The classical negation connective $\neg$ can be considered as an abbreviation: $\neg\varphi := \varphi \rightarrow \bot$. The presence of two symbols $\Box$ and $\Diamond$ for modalities is essential, because in BK^{FS} the necessity and possibility operators are not interdefinable. The set $For(\mathcal{L})$ of formulas is constructed in the standard way from the fixed set Prop of propositional variables with the help of connectives of $\mathcal{L}$.

Recall the definition of the standard translation ST_x from modal language to the first order language with the strong negation and the following set of predicate symbols: $\Sigma = \{R^2, P_1^1, P_2^1, \ldots, P_n^1, \ldots\}$. Let x and y be two different individual variables. For $\varphi \in For(\mathcal{L})$, its standard translation $ST_x(\varphi)$ is defined by induction as follows:

---

[2]Weak equivalence and weak implication are related in a usual way $\varphi \leftrightarrow \psi := (\varphi \rightarrow \psi) \wedge (\psi \rightarrow \varphi)$.

Algebraic Semantics for Fischer Servi's BK

$$ST_x(\bot) = \bot;$$
$$ST_x(p_i) = P_i(x), p_i \in Prop;$$
$$ST_x(\varphi \land \psi) = ST_x(\varphi) \land ST_x(\psi);$$
$$ST_x(\varphi \lor \psi) = ST_x(\varphi) \lor ST_x(\psi);$$
$$ST_x(\varphi \to \psi) = ST_x(\varphi) \to ST_x(\psi);$$
$$ST_x(\sim\varphi) = \sim ST_x(\varphi);$$
$$ST_x(\Box\varphi) = \forall y(R(x,y) \to ST_y(\varphi));^3$$
$$ST_x(\Diamond\varphi) = \exists y(R(x,y) \land ST_y(\varphi)).$$

The logic BK^{FS} was defined in (Odintsov & Wansing, 2017) via a standard translation into BD-FOL, Belnap-Dunn first-order logic. Following Sano and Omori (2014) we recall the semantics of BD-FOL in the language without equality and with the just defined set Σ of predicate symbols. The models of BD-FOL have the form $\mathfrak{M} = \langle M, \mu^\Sigma \rangle$, where M is a non-empty set and μ^Σ is an interpretation of symbols from Σ in M. Every unary predicate is interpreted via its extension and coextension, i.e., $\mu^\Sigma(P_i) = (P_i^+, P_i^-)$, where $P_i^+, P_i^- \subseteq M$. For R^2, we have $\mu^\Sigma(R) = (R^+, R^-)$, where $R^+, R^- \subseteq M^2$. By an *assignment in* $\mathfrak{M}$ we mean a mapping s from the set Var of individual variables into M. For a variable $x \in Var$ and two assignments s and s', we write $s \sim^x s'$ if $s(y) = s'(y)$ whenever $y \neq x$. With every model we associate two relations (verification and falsification) between assignments and formulas:

$\mathfrak{M}, s \models^+ P_i(x)$ iff $s(x) \in P_i^+$;
$\mathfrak{M}, s \models^- P_i(x)$ iff $s(x) \in P_i^-$;
$\mathfrak{M}, s \models^+ R(x,y)$ iff $(s(x), s(y)) \in R^+$;
$\mathfrak{M}, s \models^- R(x,y)$ iff $(s(x), s(y)) \in R^-$;

---

[3] To pass from $ST_x(\varphi)$ to $ST_y(\varphi)$ we simultaneously replace all occurrences of x by y and all occurrences of y by x.

$\mathfrak{M}, s \models^+ \varphi \wedge \psi$ iff $(\mathfrak{M}, s \models^+ \varphi$ and $\mathfrak{M}, s \models^+ \psi)$;
$\mathfrak{M}, s \models^- \varphi \wedge \psi$ iff $(\mathfrak{M}, s \models^- \varphi$ or $\mathfrak{M}, s \models^- \psi)$;
$\mathfrak{M}, s \models^+ \varphi \vee \psi$ iff $(\mathfrak{M}, s \models^+ \varphi$ or $\mathfrak{M}, s \models^+ \psi)$;
$\mathfrak{M}, s \models^- \varphi \vee \psi$ iff $(\mathfrak{M}, s \models^- \varphi$ and $\mathfrak{M}, s \models^- \psi)$;
$\mathfrak{M}, s \models^+ \varphi \to \psi$ iff $(\mathfrak{M}, s \models^+ \varphi$ implies $\mathfrak{M}, s \models^+ \psi)$;
$\mathfrak{M}, s \models^- \varphi \to \psi$ iff $(\mathfrak{M}, s \models^+ \varphi$ and $\mathfrak{M}, s \models^- \psi)$;
$\mathfrak{M}, s \models^+ \sim\varphi$ iff $\mathfrak{M}, s \models^- \varphi$;
$\mathfrak{M}, s \models^- \sim\varphi$ iff $\mathfrak{M}, s \models^+ \varphi$;
$\mathfrak{M}, s \not\models^+ \bot$ $\mathfrak{M}, s \models^- \bot$;

$\mathfrak{M}, s \models^+ \forall x \varphi$ iff $\forall s'(s' \sim^x s$ implies $\mathfrak{M}, s' \models^+ \varphi)$;
$\mathfrak{M}, s \models^- \forall x \varphi$ iff $\exists s'(s' \sim^x s$ and $\mathfrak{M}, s' \models^- \varphi)$;
$\mathfrak{M}, s \models^+ \exists x \varphi$ iff $\exists s'(s' \sim^x s$ and $\mathfrak{M}, s' \models^+ \varphi)$;
$\mathfrak{M}, s \models^- \exists x \varphi$ iff $\forall s'(s' \sim^x s$ implies $\mathfrak{M}, s' \models^- \varphi)$.

A formula φ is a BD-FOL-*tautology* if $\mathfrak{M}, s \models^+ \varphi$ for all $\mathfrak{M}$ and s.

The set of all $\varphi \in For(\mathcal{L})$ such that $ST_x(\varphi)$ is a BD-FOL-tautology was denoted in (Odintsov & Wansing, 2017) as BK^{FS}.

A BK^{FS}-*model* is a tuple $\mathcal{M} = \langle W, R_+, R_-, v^+, v^- \rangle$ such that W is a non-empty set, $R_+, R_- \subseteq W^2$, and $v^+, v^- : \text{Prop} \to 2^W$. Having a BK^{FS}-model $\mathcal{M} = \langle W, R_+, R_-, v^+, v^- \rangle$, we define verification and falsification relations $\models^+$ and $\models^-$ between worlds and formulas as follows:

$\mathcal{M}, w \models^+ p$ iff $w \in v^+(p)$; $\quad \mathcal{M}, w \models^- p$ iff $w \in v^-(p)$;
$\mathcal{M}, w \models^+ \varphi \wedge \psi$ iff $(\mathcal{M}, w \models^+ \varphi$ and $\mathcal{M}, w \models^+ \psi)$;
$\mathcal{M}, w \models^- \varphi \wedge \psi$ iff $(\mathcal{M}, w \models^- \varphi$ or $\mathcal{M}, w \models^- \psi)$;
$\mathcal{M}, w \models^+ \varphi \vee \psi$ iff $(\mathcal{M}, w \models^+ \varphi$ or $\mathcal{M}, w \models^+ \psi)$;
$\mathcal{M}, w \models^- \varphi \vee \psi$ iff $(\mathcal{M}, w \models^- \varphi$ and $\mathcal{M}, w \models^- \psi)$;
$\mathcal{M}, w \models^+ \varphi \to \psi$ iff $(\mathcal{M}, w \models^+ \varphi$ only if $\mathcal{M}, w \models^+ \psi)$;
$\mathcal{M}, w \models^- \varphi \to \psi$ iff $(\mathcal{M}, w \models^+ \varphi$ and $\mathcal{M}, w \models^- \psi)$;
$\mathcal{M}, w \not\models^+ \bot$; $\quad \mathcal{M}, w \models^- \bot$;
$\mathcal{M}, w \models^+ \sim\varphi$ iff $\mathcal{M}, w \models^- \varphi$;
$\mathcal{M}, w \models^- \sim\varphi$ iff $\mathcal{M}, w \models^+ \varphi$;
$\mathcal{M}, w \models^+ \Box\varphi$ iff $\forall u(w R_+ u$ only if $\mathcal{M}, u \models^+ \varphi)$;
$\mathcal{M}, w \models^- \Box\varphi$ iff $\exists u(w R_+ u$ and $\mathcal{M}, u \models^- \varphi)$.
$\mathcal{M}, w \models^+ \Diamond\varphi$ iff $\exists u(w R_+ u$ and $\mathcal{M}, u \models^+ \varphi)$;
$\mathcal{M}, w \models^- \Diamond\varphi$ iff $\forall u(w R_- u$ only if $\mathcal{M}, u \models^- \varphi)$.

Algebraic Semantics for Fischer Servi's BK

A formula $\varphi \in For(\mathcal{L})$ is $\mathsf{BK}^{\mathsf{FS}}$-*valid* if $\mathcal{M}, w \models^+ \varphi$ for all $\mathcal{M}$ and $w \in W$. For $\Gamma \cup \{\varphi\} \subseteq For(\mathcal{L})$, the relation $\Gamma \models_{\mathsf{BK}^{\mathsf{FS}}} \varphi$ holds iff for every $\mathsf{BK}^{\mathsf{FS}}$-model $\mathcal{M}$ and its world w,

$$\mathcal{M}, w \models^+ \varphi \text{ whenever } \mathcal{M}, w \models^+ \psi \text{ for all } \psi \in \Gamma.$$

Odintsov and Wansing (2017) established the equivalence of Kripke style and first order semantics for $\mathsf{BK}^{\mathsf{FS}}$.

Theorem 1 *For* $\varphi \in For(\mathcal{L}^m)$, $\varphi \in \mathsf{BK}^{\mathsf{FS}}$ *iff* φ *is* $\mathsf{BK}^{\mathsf{FS}}$-*valid*.

To characterize $\mathsf{BK}^{\mathsf{FS}}$ syntactically, we consider three groups of axioms:

I) axioms of classical propositional logic in the language $\{\vee, \wedge, \rightarrow, \bot\}$;

II) strong negation axioms:

$$\sim(p \wedge q) \leftrightarrow (\sim p \vee \sim q), \quad \sim(p \rightarrow q) \leftrightarrow (p \wedge \sim q),$$
$$\sim(p \vee q) \leftrightarrow (\sim p \wedge \sim q), \quad \sim\sim p \leftrightarrow p, \quad \text{and} \quad \sim\bot;$$

III) modal axioms:

$$\Box(p \rightarrow q) \rightarrow (\Box p \rightarrow \Box q), \quad \neg\sim\Box p \leftrightarrow \Box\neg\sim p,$$
$$\sim\Diamond(p \vee q) \leftrightarrow (\sim\Diamond p \wedge \sim\Diamond q), \quad \neg\Diamond p \leftrightarrow \Box\neg p;$$

together with the following rules:

$$(Sub) \; \frac{\varphi(p_1, \ldots p_n)}{\varphi(\psi_1, \ldots \psi_n)}, \quad (MP) \; \frac{\varphi \quad \varphi \rightarrow \psi}{\psi}, \quad (RN) \; \frac{\varphi}{\Box\varphi},$$

$$(RM^{\sim\Diamond}) \; \frac{\sim p \rightarrow \sim q}{\sim\Diamond p \rightarrow \sim\Diamond q}.$$

The set $\mathsf{Th}(\mathsf{BK}^{\mathsf{FS}})$ of $\mathsf{BK}^{\mathsf{FS}}$-*theorems* is defined as the least set containing the above axioms and closed under the rules Sub, MP, RN, and $RM^{\sim\Diamond}$. For $\Gamma \cup \{\varphi\} \subseteq For(\mathcal{L})$, the relation $\Gamma \vdash_{\mathsf{BK}^{\mathsf{FS}}} \varphi$ holds iff φ can be obtained from elements of $\Gamma \cup \mathsf{Th}(\mathsf{BK}^{\mathsf{FS}})$ with the help of MP only.

The strong completeness theorem for $\mathsf{BK}^{\mathsf{FS}}$ looks as follows.

Theorem 2 (Odintsov & Wansing, 2017) *For any* $\Gamma \cup \{\varphi\} \subseteq For\mathcal{L}$, *the following equivalence holds:*

$$\Gamma \vdash_{\mathsf{BK}^{\mathsf{FS}}} \varphi \quad \text{iff} \quad \Gamma \models_{\mathsf{BK}^{\mathsf{FS}}} \varphi.$$

In particular, $\mathsf{Th}(\mathsf{BK}^{\mathsf{FS}}) = \mathsf{BK}^{\mathsf{FS}}$.

The consequence relation $\vdash_{\mathsf{BK^{FS}}}$ as we can see from the above theorem preserves the validity in a world of a model. Due to this reason the relation $\vdash_{\mathsf{BK^{FS}}}$ defined via the rule MP is called a local consequence relation. The global consequence relation $\vdash^*_{\mathsf{BK^{FS}}}$ is defined via all rules. More exactly, for $\Gamma \cup \{\varphi\} \subseteq For(\mathcal{L})$, the relation $\Gamma \vdash^*_{\mathsf{BK^{FS}}} \varphi$ holds iff φ can be obtained from elements of $\Gamma \cup \mathsf{BK^{FS}}$ with the help of MP, RN, and $RM^{\sim\Diamond}$.

3 Fs-twist structures

In this section we characterize the consequence relation $\vdash^*_{\mathsf{BK^{FS}}}$ via twist structures over bimodal algebras.

An algebraic system $\mathcal{A} = \langle A, \vee, \wedge, \neg, \Box_+, \Box_- \rangle$ is a bimodal algebra if its $\{\Box_+, \Box_-\}$-free reduct $\langle A, \vee, \wedge, \neg \rangle$ is a Boolean algebra and both modal operators satisfy for all $a, b \in A$ the following conditions: 1) $\Box_\varepsilon 1 = 1$, 2) $\Box_\varepsilon(a \wedge b) = \Box_\varepsilon a \wedge \Box_\varepsilon b$, where $\varepsilon \in \{+, -\}$ and $1 = a \vee \neg a$ is the greatest element of Boolean algebra $\langle A, \vee, \wedge, \neg \rangle$. We use the abbreviations

$$0 := \neg 1, \quad \Diamond_\varepsilon a := \neg \Box_\varepsilon \neg a, \quad \varepsilon \in \{+, -\}.$$

Recall that 0 is the least element of $\langle A, \vee, \wedge, \neg \rangle$, $\Diamond_\varepsilon 0 = 0$, and $\Diamond_\varepsilon(a \vee b) = \Diamond_\varepsilon a \vee \Diamond_\varepsilon b$ for $\varepsilon \in \{+, -\}$ and $a, b \in A$.

Definition 1 Let $\mathcal{A} = \langle A, \vee, \wedge, \neg, \Box_+, \Box_- \rangle$ be a bimodal algebra.

1. *The full fs-twist structure over $\mathcal{A}$ is an algebra*

$$\mathcal{A}^{\bowtie}_{\mathsf{fs}} = \langle A \times A, \vee, \wedge, \to, \bot, \sim, \Box, \Diamond \rangle$$

with twist operations:

$$(a,b) \wedge (c,d) = (a \wedge c, b \vee d), \quad (a,b) \vee (c,d) = (a \vee c, b \wedge d),$$

$$(a,b) \to (c,d) = (\neg a \vee c, a \wedge d), \quad \bot = (0,1), \quad \sim(a,b) = (b,a),$$

$$\Box(a,b) = (\Box_+ a, \Diamond_+ b), \quad \Diamond(a,b) = (\Diamond_+ a, \Box_- b).$$

2. *An fs-twist structure over a bimodal algebra $\mathcal{A}$ is a subalgebra $\mathcal{B}$ of $\mathcal{A}^{\bowtie}_{\mathsf{fs}}$ such that its projection $\pi_1 \mathcal{B}$ onto the first coordinate is A.*

Let $\mathcal{B}$ be an fs-twist structure over a bimodal algebra. A homomorphism from the algebra of formulas $For(\mathcal{L})$ to $\mathcal{B}$ is called a $\mathcal{B}$-valuation.

Algebraic Semantics for Fischer Servi's BK

For $\Gamma \cup \{\varphi\} \subseteq For(\mathcal{L})$, the relation $\Gamma \models_\mathcal{B} \varphi$ means that for every $\mathcal{B}$-valuation v, if $\pi_1 v(\psi) = 1$ for all $\psi \in \Gamma$, then $\pi_1 v(\varphi) = 1$. We write $\Gamma \models_{FS}^{\bowtie} \varphi$ iff $\Gamma \models_\mathcal{B} \varphi$ for every fs-twist structure $\mathcal{B}$ over a bimodal algebra.

A set $\Gamma \subseteq For(\mathcal{L})$ is said to be *non-trivial w.r.t.* $\mathsf{BK}^{\mathsf{FS}}$, if there is a formula φ such that $\Gamma \nvdash^*_{\mathsf{BK}^{\mathsf{FS}}} \varphi$. With an arbitrary set of formulas Γ, non-trivial w.r.t. $\mathsf{BK}^{\mathsf{FS}}$, we associate the equivalence relation $\equiv_\Gamma$ such that for any $\varphi, \psi \in For(\mathcal{L})$, $\varphi \equiv_\Gamma \psi$ iff $\Gamma \vdash^*_{\mathsf{BK}^{\mathsf{FS}}} \varphi \leftrightarrow \psi$.

The equivalence class of φ wrt $\equiv_\Gamma$ is denoted as $[\varphi]_\Gamma$, and the family of all equivalence classes as M_Γ.

Proposition 1 *Let* $\Gamma \cup \{\varphi, \varphi', \psi, \psi'\} \subseteq For(\mathcal{L})$.

1. *If* $\varphi \equiv_\Gamma \varphi'$ *and* $\psi \equiv_\Gamma \psi'$, *then*

$$\varphi \wedge \psi \equiv_\Gamma \varphi' \wedge \psi', \quad \varphi \vee \psi \equiv_\Gamma \varphi' \vee \psi', \quad \neg\varphi \equiv_\Gamma \neg\varphi', \quad \Box\varphi \equiv_\Gamma \Box\varphi'.$$

2. *If* $\varphi \equiv_\Gamma \varphi'$, *then* $\sim\Diamond\sim\varphi \equiv_\Gamma \sim\Diamond\sim\varphi'$.

Proof. 1. The fact that $\equiv_\Gamma$ has congruence properties w.r.t. $\wedge, \vee, \neg,$ and $\Box$ can be proved in the same way as for a similar equivalence relation defined over the least normal modal logic K, since our axiomatization of $\mathsf{BK}^{\mathsf{FS}}$ contains all axioms and rules from the standard axiomatics of K.

2. From $\Gamma \vdash^*_{\mathsf{BK}^{\mathsf{FS}}} \varphi \leftrightarrow \varphi'$ we pass to $\Gamma \vdash^*_{\mathsf{BK}^{\mathsf{FS}}} \sim\sim\varphi \leftrightarrow \sim\sim\varphi'$ using instances of the axiom $p \leftrightarrow \sim\sim p$ and the transitivity of $\leftrightarrow$. From $\Gamma \vdash^*_{\mathsf{BK}^{\mathsf{FS}}} \sim\sim\varphi \leftrightarrow \sim\sim\varphi'$ we obtain $\Gamma \vdash^*_{\mathsf{BK}^{\mathsf{FS}}} \sim\Diamond\sim\varphi \leftrightarrow \sim\Diamond\sim\varphi'$ by $RM^{\sim\Diamond}$. $\square$

Due to this proposition the operations $\vee, \wedge, \neg, \Box_+, \Box_-$ on M_Γ given by the rules:

$$\begin{aligned}
[\varphi]_\Gamma * [\psi]_\Gamma &:= [\varphi * \psi]_\Gamma, \quad \text{where } * \in \{\vee, \wedge\}; \\
\neg[\varphi]_\Gamma &:= [\neg\varphi]_\Gamma; \\
\Box_+[\varphi]_\Gamma &:= [\Box\varphi]_\Gamma; \\
\Box_-[\varphi]_\Gamma &:= [\sim\Diamond\sim\varphi]_\Gamma;
\end{aligned}$$

are well defined. Moreover, we have the following.

Proposition 2 $\mathcal{M}_\Gamma := \langle M_\Gamma, \vee, \wedge, \neg, \Box_+, \Box_- \rangle$ *is a bimodal algebra.*

Proof. $\langle M_\Gamma, \vee, \wedge, \neg \rangle$ is a Boolean algebra by the axioms of classical propositional logic. The required properties of $\Box_+$ and $\Box_-$ follow from the fact that formulas

$$\Box(\varphi \to \varphi), \quad \sim\Diamond\sim(\varphi \to \varphi), \quad \Box(\varphi \wedge \psi) \leftrightarrow (\Box\varphi \wedge \Box\psi),$$

$$\sim\Diamond\sim(\varphi \wedge \psi) \leftrightarrow (\sim\Diamond\sim\varphi \wedge \sim\Diamond\sim\psi)$$

belong to $\mathsf{BK}^{\mathsf{FS}}$, which can be checked with the help of Theorem 2. □

Consider the full fs-twist structure $(\mathcal{M}_\Gamma)_{\mathsf{fs}}^{\bowtie}$ over $\mathcal{M}_\Gamma$.

Proposition 3 *The subset*

$$\mathsf{FS}_\Gamma^{\bowtie} = \{([\varphi]_\Gamma, [\sim\varphi]_\Gamma) \mid \varphi \in For(\mathcal{L})\}$$

of $|(\mathcal{M}_\Gamma)_{\mathsf{fs}}^{\bowtie}|$ *is closed under twist operations.*

Proof. The group of strong negation axioms contains De Morgan laws, which easily imply that $\mathsf{FS}_\Gamma^{\bowtie}$ is closed under $\vee$ and $\wedge$. Further, the axiom $\sim(p \to q) \leftrightarrow (p \wedge \sim q)$ entails that $\mathsf{FS}_\Gamma^{\bowtie}$ is closed under $\to$. $\mathsf{FS}_\Gamma^{\bowtie}$ is closed under strong negation due to $p \leftrightarrow \sim\sim p$. Indeed,

$$\sim([\varphi]_\Gamma, [\sim\varphi]_\Gamma) = ([\sim\varphi]_\Gamma, [\varphi]_\Gamma) = ([\sim\varphi]_\Gamma, [\sim\sim\varphi]_\Gamma).$$

In what follows we write $[\varphi]$ instead of $[\varphi]_\Gamma$. For $\Box$, we have

$$\Box([\varphi], [\sim\varphi]) = (\Box_+[\varphi], \neg\Box_+\neg[\sim\varphi]) = ([\Box\varphi], [\neg\Box\neg\sim\varphi]) =$$
$$= ([\Box\varphi], [\neg\neg\sim\Box\varphi]) = ([\Box\varphi], [\sim\Box\varphi]).$$

The third equality is due to axiom $\neg\sim\Box p \leftrightarrow \Box\neg\sim p$ and Item 1 of Proposition 1.

Finally, consider the $\Diamond$-operator.

$$\Diamond([\varphi], [\sim\varphi]) = (\neg\Box_+\neg[\varphi], \Box_-[\sim\varphi]) = ([\neg\Box\neg\varphi], [\sim\Diamond\sim\sim\varphi]) =$$
$$= ([\neg\neg\Diamond\varphi], [\sim\Diamond\varphi]) = ([\Diamond\varphi], [\sim\Diamond\varphi]).$$

The third equality is due to axiom $\neg\Diamond p \leftrightarrow \Box\neg p$ and $\sim\Diamond\sim\sim\varphi \leftrightarrow \sim\Diamond\varphi \in \mathsf{BK}^{\mathsf{FS}}$. The last fact can be checked with the help of Theorem 2. □

Thus, we can consider the fs-twist stricture $\mathcal{FS}_\Gamma^{\bowtie}$ over $\mathcal{M}_\Gamma$ with the support $\mathsf{FS}_\Gamma^{\bowtie}$ and prove the strong completeness theorem for $\vdash^*_{\mathsf{BK}^{\mathsf{FS}}}$.

Theorem 3 *Let Γ be a set of formulas non-trivial w.r.t. $\mathsf{BK}^{\mathsf{FS}}$, and let φ be a formula. The following conditions are equivalent:*

1) $\Gamma \vdash^*_{\mathsf{BK}^{\mathsf{FS}}} \varphi$; 2) $\Gamma \models^{\bowtie}_{\mathsf{FS}} \varphi$; 3) $\Gamma \models_{\mathcal{FS}_\Gamma^{\bowtie}} \varphi$.

Proof. 1)⇒2) This implication can be established in the standard way by induction on the length of proof.

2)⇒3) It is trivial.

3)⇒1) Assume that $\Gamma \not\vdash^*_{\mathsf{BK^{FS}}} \varphi$. In this case, $[\varphi]_\Gamma$ is different from the greatest element $1_{\mathcal{M}_\Gamma}$ of $\mathcal{M}_\Gamma$, which is equal to the equivalence class consisting of formulas inferable from Γ in $\mathsf{BK^{FS}}$.

Let us consider the twist-structure $\mathcal{FS}^{\bowtie}_\Gamma$ and its valuation v given by the rule $v(p) = ([p]_\Gamma, [\sim p]_\Gamma)$. It is easy to prove by induction on the structure of formulas that
$$v(\psi) = ([\psi]_\Gamma, [\sim\psi]_\Gamma)$$
for any formula ψ. In particular, $v(\varphi) = ([\varphi]_\Gamma, [\sim\varphi]_\Gamma)$ holds. Therefore, $\pi_1 v(\varphi) \neq 1_{\mathcal{M}_\Gamma}$. At the same time, for every $\psi \in \Gamma$, we have $\pi_1 v(\psi) = 1_{\mathcal{M}_\Gamma}$ since, obviously, $\Gamma \vdash^*_{\mathsf{BK^{FS}}} \psi$. Thus, $\Gamma \not\models_{\mathcal{FS}^{\bowtie}_\Gamma} \varphi$. □

4 Algebraizability

The aim of this section is to prove that the abstract closure of the class of fs-twist structures over bimodal algebras forms a variety and that this variety provides an equivalent algebraic semantics for $\mathsf{BK^{FS}}$ in the sense of Blok and Pigozzi (1989), i.e., that $\mathsf{BK^{FS}}$ is algebraizable.

Definition 2 *An algebra* $\mathcal{A} = \langle A, \vee, \wedge, \to, \bot, \sim, \Box, \Diamond \rangle$ *is an FS-lattice, if* $\langle A, \vee, \wedge, \bot, \sim\bot \rangle$ *is a bounded distributive lattice and the following conditions hold for all* $a, b \in A$:

1. $\sim\sim a = a$, $\sim(a \vee b) = \sim a \wedge \sim b$, $\sim(a \wedge b) = \sim a \vee \sim b$;[4]

2. $\neg(a \vee b) = \neg a \wedge \neg b$, $\neg(a \wedge b) = \neg a \vee \neg b$, $\neg\bot = \sim\bot$, $\neg\sim\bot = \bot$,[5] where $\neg a = a \to \bot$.

3. $\neg a \wedge \neg\neg a = \bot$, $\neg a \vee \neg\neg a = \sim\bot$;

4. $a \to b = \neg a \vee b$;

5. $\sim\neg a = \neg\neg a$;

6. $\Box\neg a = \neg\Diamond a$, $\Diamond\neg a = \neg\Box a$, $\neg\sim\Box a = \Box\neg\sim a$, $\neg\sim\Diamond a = \neg\Diamond\neg\sim a$;

7. $\Box\sim\bot = \sim\bot$, $\Box(\neg a \wedge \neg b) = \Box\neg a \wedge \Box\neg b$;

---

[4] So $\langle A, \vee, \wedge, \sim \rangle$ is a De Morgan algebra.
[5] In this way, $\langle A, \vee, \wedge, \neg, \bot, \sim\bot \rangle$ is an Ockham lattice (see Urquhart, 1979).

8. $\sim\lozenge\bot = \sim\bot$, $\sim\lozenge\sim(\neg a \wedge \neg b) = \sim\lozenge\sim\neg a \wedge \sim\lozenge\sim\neg b$;

9. if $\neg\neg a = \neg b$ and $\neg\sim a = \neg\sim b$, then $a = b$.

Remark 1 Identity $\neg\neg\neg a = \neg a$ holds on FS-lattices. Indeed, by Item 3 of the previous definition both $\neg\neg\neg a$ and $\neg a$ are complements of $\neg\neg a$. An element of a distributive lattice may have at most one complement, consequently, $\neg\neg\neg a = \neg a$.

Proposition 4 *Every twist-structure $\mathcal{A}$ over a bimodal algebra $\mathcal{B}$ is an FS-lattice.*

Proof. By a direct computation one can check that $\mathcal{A}$ is a bounded distributive lattice with the least element $\bot$ and the greatest element $\sim\bot = (1, 0)$. Items 1–8 of Definition 2 also can be checked via a routine computation. Let us check Item 9.

Assume that $\neg(a, b) = \neg(c, d)$ and $\neg\sim(a, b) = \neg\sim(c, d)$ for (a, b) and (c, d) from $\mathcal{A}$, i.e., we have

$$(\neg a, a) = (\neg c, c) \text{ and } (\neg b, b) = (\neg d, d).$$

The first equation implies $a = c$ and the second one implies $b = d$. Thus, $(a, b) = (c, d)$. □

Remark 2 It follows from the definition of twist operations $\vee$ and $\wedge$ that the lattice order $\leq_{\bowtie}$ on $\mathcal{A}$ is such that

$$(a, b) \leq_{\bowtie} (c, d) \text{ iff } a \leq c \text{ and } d \leq b,$$

where $\leq$ is the lattice order of $\mathcal{B}$.

Now we show that the class of FS-lattices coincides with the abstract closure of the class of fs-twist structures over bimodal algebras, i.e., that every FS-lattice is isomorphic to a twist-structure over a bimodal algebra.

Proposition 5 *For every FS-lattice $\mathcal{A} = \langle A, \vee, \wedge, \rightarrow, \bot, \sim, \square, \lozenge \rangle$, the set $A^{\neg} = \{\neg a \mid a \in \mathcal{A}\}$ is closed under the operations $\vee, \wedge, \neg, \square,$ and $\square_{-}$, where $\square_{-} a := \neg\neg\sim\lozenge\sim a$ for $a \in A^{\neg}$. Moreover, the algebra*

$$\mathcal{A}_{\bowtie} := \langle A^{\neg}, \vee, \wedge, \neg, \square, \square_{-} \rangle$$

is a bimodal algebra.

Algebraic Semantics for Fischer Servi's **BK**

Proof. The fact that $A^{\neg}$ is closed under $\neg$ is trivial. De Morgan laws for $\neg$-connective (Item 2)[6] imply that $A^{\neg}$ is closed under the lattice operations. We have $\bot, {\sim}\bot \in A^{\neg}$, since $\bot = \neg{\sim}\bot$ and ${\sim}\bot = \neg\bot$ according to Item 2. Thus, $\langle A^{\neg}, \vee, \wedge, \bot, {\sim}\bot \rangle$ is a bounded distributive lattice. The element $\neg a$ is a complement of $a \in A^{\neg}$ by Item 3, consequently, $\langle A^{\neg}, \vee, \wedge, \neg \rangle$ is a Boolean algebra. Item 6 implies that $A^{\neg}$ is closed under $\Box$, and Item 7 means that $\Box$ is a necessity operator on $\mathcal{A}_{\bowtie}$. That $A^{\neg}$ is closed under $\Box_-$ follows from the definition of $\Box_-$. Let us check that $\Box_-$ is a necessity operator on $\mathcal{A}_{\bowtie}$. Using identities ${\sim}{\sim}a = a$ (Item 1), ${\sim}\Diamond\bot = {\sim}\bot$ (Item 8), $\neg{\sim}\bot = \bot$ and $\neg\bot = {\sim}\bot$ (Item 2) yield

$$\Box_-{\sim}\bot = \neg\neg{\sim}\Diamond{\sim}{\sim}\bot = \neg\neg{\sim}\Diamond\bot = \neg\neg{\sim}\bot = \neg\bot = {\sim}\bot.$$

Finally, applying Item 8 and De Morgan law for $\neg$ and $\wedge$ (Item 2) we obtain

$$\Box_-(\neg a \wedge \neg b) = \neg\neg{\sim}\Diamond{\sim}(\neg a \wedge \neg b) = \neg\neg({\sim}\Diamond{\sim}\neg a \wedge {\sim}\Diamond{\sim}\neg b) =$$

$$= \neg\neg{\sim}\Diamond{\sim}\neg a \wedge \neg\neg{\sim}\Diamond{\sim}\neg b = \Box_-\neg a \wedge \Box_-\neg b.$$

We have thus proved that $\mathcal{A}_{\bowtie} := \langle A^{\neg}, \vee, \wedge, \neg, \Box, \Box_- \rangle$ is a bimodal algebra. $\square$

In fact, we have just distinguished in an FS-lattice $\mathcal{A}$ the underlying algebra of an fs-twist structure, to which $\mathcal{A}$ is isomorphic.

Proposition 6 *Every* FS-*lattice $\mathcal{A}$ is isomorphic to an fs-twist structure over bimodal algebra $\mathcal{A}_{\bowtie}$.*

Proof. It is enough to embed $\mathcal{A}$ into the full fs-twist structure $(\mathcal{A}_{\bowtie})^{\bowtie}_{\text{fs}}$. To this end we consider a mapping $h: \mathcal{A} \to (\mathcal{A}_{\bowtie})^{\bowtie}_{\text{fs}}$ given by the rule $h(a) := (\neg\neg a, \neg\neg{\sim}a)$.

De Morgan laws for both negations ${\sim}$ and $\neg$ allow us to check that h preserves lattice operations. From ${\sim}{\sim}a = a$ (Item 1) we infer

$$h({\sim}a) = (\neg\neg{\sim}a, \neg\neg{\sim}{\sim}a) = (\neg\neg{\sim}a, \neg\neg a) = {\sim}h(a).$$

Applying ${\sim}{\sim}a = a$ together with ${\sim}\neg a = \neg\neg a$ (Item 5) we obtain

$$h(\neg a) = (\neg\neg\neg a, \neg\neg{\sim}\neg a) = (\neg\neg\neg a, \neg\neg\neg\neg a) = (\neg\neg\neg a, {\sim}{\sim}\neg\neg a) =$$
$$= (\neg\neg\neg a, \neg\neg a) = \neg(\neg\neg a, \neg\neg{\sim}a) = \neg h(a).$$

---

[6]Here and up to the end of this section "Item n" means "Item n of Definition 2".

According to Item 4 $\to$ can be expressed via $\neg$ and $\lor$, consequently, h preserves $\to$ too. Since $\bot$ and $\sim\!\bot$ are the least and the greatest elements of $\mathcal{A}_{\bowtie}$, the pair $(\bot, \sim\!\bot)$ is the least element of $(\mathcal{A}_{\bowtie})^{\bowtie}_{fs}$, i.e., $\bot_{(\mathcal{A}_{\bowtie})^{\bowtie}_{fs}} = (\bot, \sim\!\bot)$. By Item 2 we have

$$h(\bot) = (\neg\neg\bot, \neg\neg\sim\!\bot) = (\neg\sim\!\bot, \neg\bot) = (\bot, \sim\!\bot).$$

It follows from Item 6 and the identity $\neg\neg\neg a = \neg a$ (see Remark 1) that

$$h(\Box a) = (\neg\neg\Box a, \neg\neg\sim\!\Box a) = (\neg\Diamond\neg a, \neg\Box\neg\sim\!a) = $$
$$= (\Box\neg\neg a, \neg\Box\neg\neg\neg\sim\!a) = \Box(\neg\neg a, \neg\neg\sim\!a) = \Box h(a).$$

Finally, using again Item 6, $\neg\neg\neg a = \neg a$, and $\sim\!\neg a = \neg\neg a$ (Item 5) we obtain

$$h(\Diamond a) = (\neg\neg\Diamond a, \neg\neg\sim\!\Diamond a) = (\neg\Box\neg a, \neg\neg\sim\!\Diamond\neg\sim\!a) =$$
$$= (\neg\Box\neg a, \neg\neg\sim\!\Diamond\neg\neg\neg\sim\!a) = (\neg\Box\neg\neg\neg a, \neg\neg\sim\!\Diamond\sim\!\neg\neg\neg\sim\!a) =$$
$$= \Diamond(\neg\neg a, \neg\neg\sim\!a) = \Diamond h(a).$$

We have thus proved that h is a homomorphism. It remains to check that h is one-to-one. If $h(a) = h(b)$, then $\neg\neg a = \neg\neg b$ and $\neg\neg\sim\!a = \neg\neg\sim\!b$. Obviously, we also have $\neg\neg\neg a = \neg\neg\neg b$ and $\neg\neg\neg\sim\!a = \neg\neg\neg\sim\!b$. Using $\sim\!\neg a = \neg\neg a$ yields $\sim\!\sim\!\neg a = \sim\!\sim\!\neg b$ and $\sim\!\sim\!\neg\sim\!a = \sim\!\sim\!\neg\sim\!b$, from which by deleting double strong negations we obtain $\neg a = \neg b$ and $\neg\sim\!a = \neg\sim\!b$. Item 9 now implies $a = b$.

We have $\pi_1(h(\neg a)) = \neg\neg\neg a = \neg a$. Consequently, $\pi_1 h(\mathcal{A}) = \mathcal{A}_{\bowtie}$ and $h(\mathcal{A})$ is an fs-twist structure over $\mathcal{A}_{\bowtie}$. $\square$

It follows from the definition that the class of all FS-lattices is a quasi-variety. It turns out that it is even a variety.

Theorem 4 *The class $\mathcal{V}^{FS}$ of all FS-lattices forms a variety.*

Proof. To prove this statement we have to replace Item 9 of Definition 2, which has the form of quasi-identity, by several identities, namely, by the following two identities:

I) $\neg a \lor a = \neg\sim\!a \lor \sim\!a$;

II) $a \land (\neg a \lor b) \leq b \lor (\sim\!\neg\neg b \land a)$.

Algebraic Semantics for Fischer Servi's BK

An algebraic system $\mathcal{A} = \langle A, \vee, \wedge, \rightarrow, \bot, \sim, \Box, \Diamond \rangle$ is an **FS**-lattice if and only if $\mathcal{A}$ satisfies the identities of bounded distributive lattices with the least element $\bot$ and the greatest element $\sim\!\bot$, the identities of Items 1–8 from Definition 2, and the identities I) and II).

This equivalence can be obtained by rewording the proof of Theorem 4.2 of Odintsov and Latkin (2012), which states that the class of **BK**-lattices forms a variety. **BK**-lattices have another $\Diamond$-modality, which is dual to $\Box$ via strong negation. The latter means that **BK**-lattices satisfy the identity $\Diamond a = \sim\!\Box\!\sim\!a$, i.e. they can be defined in the language without $\Diamond$-modality. The analysis of the proof of Theorem 4.2 (Odintsov & Latkin, 2012) shows that it only uses the identities of bounded distributive lattices, the identities of Items 1–4, and the identities I) and II). So it also works in our case. □

Recall that an *equation* in the language $\mathcal{L}$ ($\mathcal{L}$-*equation*) is a formal expression of the form $\varphi \simeq \psi$, where φ and ψ are in $For(\mathcal{L})$.

Let $\mathcal{A} = \langle A, \vee, \wedge, \rightarrow, \bot, \sim, \Box, \Diamond \rangle$ be an algebraic system. For a set Γ of $\mathcal{L}$-equations and an $\mathcal{L}$-equation $\varphi \simeq \psi$, the relation $\Gamma \models_\mathcal{A} \varphi \simeq \psi$ holds iff for every $\mathcal{A}$-valuation v, we have $v(\varphi) = v(\psi)$ whenever $v(\chi) = v(\eta)$ for all $\chi \simeq \eta \in \Gamma$. For a class K of algebraic systems, the relation $\Gamma \models_K \varphi \simeq \psi$ holds iff $\Gamma \models_\mathcal{A} \varphi \simeq \psi$ for all $\mathcal{A} \in K$.

Now we are ready to prove that the global consequence relation $\vdash^*_{\mathsf{BK^{FS}}}$ of the logic $\mathsf{BK^{FS}}$ is algebraizable, i.e., that $\vdash^*_{\mathsf{BK^{FS}}}$ has an equivalent algebraic semantics (for details see Blok & Pigozzi, 1989).

Theorem 5 *The variety $\mathcal{V}^{\mathsf{FS}}$ provides an equivalent algebraic semantics for the consequence relation $\vdash^*_{\mathsf{BK^{FS}}}$ with defining equation $\neg p = \bot$ and equivalence formula $p \Leftrightarrow q$.*

Proof. First we prove that $\mathcal{V}^{\mathsf{FS}}$ provides an equivalent algebraic semantics for the consequence relation $\vdash^*_{\mathsf{BK^{FS}}}$ with defining equation $\neg p = \bot$, i.e., that for every $\Gamma \cup \{\varphi\} \subseteq For(\mathcal{L})$, the following equivalence holds:

$$\Gamma \vdash^*_{\mathsf{BK^{FS}}} \varphi \text{ iff } \{\neg\psi = \bot \mid \psi \in \Gamma\} \models_{\mathcal{V}^{\mathsf{FS}}} \neg\varphi = \bot. \tag{1}$$

According to Theorem 3 $\Gamma \vdash^*_{\mathsf{BK^{FS}}} \varphi$ is equivalent to $\Gamma \models^{\bowtie}_{\mathsf{FS}} \varphi$, which means that for every fs-twist structure $\mathcal{A}$ over a bimodal algebra, for every $\mathcal{A}$-valuation v, $\pi_1 v(\varphi) = 1$ whenever $\pi_1 v(\psi) = 1$ for all $\psi \in \Gamma$. This fact and Proposition 6 imply that for every $\Gamma \cup \{\varphi\} \subseteq For(\mathcal{L})$, the equivalence (1) holds if and only if for every fs-twist structure $\mathcal{A}$ over a bimodal algebra,

$$\Gamma \models^{\bowtie}_{\mathcal{A}} \varphi \text{ iff } \{\neg\psi = \bot \mid \psi \in \Gamma\} \models_\mathcal{A} \neg\varphi = \bot. \tag{2}$$

The equivalence (2) in its own right readily follows from the following simple lemma.

Lemma 1 *Let $\mathcal{A}$ be an fs-twist structure over a bimodal algebra, v an $\mathcal{A}$-valuation, and $\varphi \in For(\mathcal{L})$. Then*

$$\pi_1 v(\varphi) = 1 \text{ iff } v(\neg\varphi) = \bot.$$

Proof. Let $v(\varphi) = (a,b)$. Then $v(\neg\varphi) = \neg v(\varphi) = (\neg a, a)$. It is clear that $\pi_1 v(\varphi) = 1$ iff $(\neg a, a) = (0,1)$. □

To prove that the algebraic semantics for $\vdash^*_{BK^{FS}}$ provided by $\mathcal{V}^{FS}$ is equivalent with equivalence formula $p \Leftrightarrow q$ we have to show that for every equation $\varphi \simeq \psi$,

$$\varphi \simeq \psi =\models_{\mathcal{V}^{FS}} \neg(\varphi \Leftrightarrow \psi) = \bot,$$

where $\varphi \simeq \psi =\models_K \chi \simeq \eta$ means that $\varphi \simeq \psi \models_K \chi \simeq \eta$ and $\chi \simeq \eta \models_K \varphi \simeq \psi$. In view of Proposition 6 it would be enough to check that for every fs-twist structure $\mathcal{A}$ over a bimodal algebra,

$$\varphi \simeq \psi =\models_{\mathcal{A}} \neg(\varphi \Leftrightarrow \psi) = \bot.$$

Let v be an $\mathcal{A}$-valuation such that $v(\varphi) = (a,b)$ and $v(\psi) = (c,d)$. The direct computation shows that

$$v(\neg(\varphi \Leftrightarrow \psi)) = (\neg((a \leftrightarrow c) \wedge (b \leftrightarrow d)), (a \leftrightarrow c) \wedge (b \leftrightarrow d)).$$

Thus, $v(\neg(\varphi \Leftrightarrow \psi)) = (0,1)$ iff $(a \leftrightarrow c) \wedge (b \leftrightarrow d) = 1$ iff $a = c$ and $b = d$ iff $v(\varphi) = v(\psi)$. □

We have thus proved that the consequence relation $\vdash^*_{BK^{FS}}$ is algebraizable. Despite unusual interaction between $\Box$ and $\Diamond$ modalities, the Fischer Servi's version of Belnapian modal logic admits standard semantical treatment.

References

Blok, W. J., & Pigozzi, D. (1989). *Algebraizable Logics*. Providence: A.M.S.

Fischer Servi, G. (1984). Axiomatizations for some intuitionistic modal logics. *Red. Sem. Mat. Universi. Politec. Torino, 42*, 179–194.

Grefe, C. (1998). Fischer Servi's intuitionistic modal logic has the finite modal property. In M. Kracht, M. de Rijke, H. Wansing, & M. Zakharyaschev (Eds.), *Advances in Modal Logic* (Vol. 2, pp. 85–98). CSLI Publications.

Odintsov, S., & Latkin, E. (2012). **BK**-lattices. algebraic semantics for belnapian modal logics. *Studia Logica, 100*, 319–338.

Odintsov, S., & Wansing, H. (2010). Modal logics with Belnapian truth values. *Journal of Applied Non-Classical Logics, 20*, 279–301.

Odintsov, S., & Wansing, H. (2017). Disentangling fde-based paraconsistent modal logics. *Studia Logica, 105*, 1221–1254.

Sano, K., & Omori, H. (2014). An expansion of first-order Belnap-Dunn logic. *Logic Journal of the IGPL, 22*, 458–481.

Simpson, A. K. (1994). *The Proof Theory and Semantics of Intuitionistic Modal Logic* (Unpublished doctoral dissertation). University of Edinburgh, Edinburgh.

Urquhart, A. (1979). Distributive lattices with a dual homomorphic operation. *Studia Logica, 38*, 201–209.

Sergei Odintsov
Sobolev Institute of Mathematics
Novosibirsk, Russia
E-mail: odintsov@math.nsc.ru

Logic as a (Natural) Science

JAROSLAV PEREGRIN[1]

Abstract: The thesis that logic is a science is not too controversial – logic seems to be so close to mathematics that that its allegiance to science seems to be obvious. In this paper I argue that though logic is a science (or something close to science), it is not because it would be akin to mathematics; that it is much closer to a natural science like physics, for just like physics it accounts for some part of reality (in case of logic it is the rules of reasoning as an ongoing human activity) largely in terms of mathematical models. In the paper I consider some recent proposal for seeing logic as a theory of reasoning (T. Williamson, G. Russell) and I try to amend what I see as their shortcomings.

Keywords: logical laws, excluded middle, modus ponens, nature of logic, logical validity, reasoning

1 Is logic a science?

Logic was born from the rib of philosophy. It is true that, from the outset, its position within the network of philosophical disciplines was somewhat peculiar: unlike ontology or epistemology, logic was more of a tool ("organon") of philosophy than directly part of its subject matter. On the other hand, the link of logic to philosophy was, and was to remain for many centuries, rather firm – neither mathematicians, nor other scientists paid much attention to it.

It was only in the second half of the nineteenth century that logic started to attract mathematicians. Boole, Frege, Peano, etc., the ur-fathers of modern logic, were all mathematicians, which helped them raise logic to a new level of rigor. On the one hand, they wanted to employ logic to fortify the foundations of mathematics (especially Frege and Peano) and, on the other, they wanted, the other way around, use mathematics to advance logic, which they saw as an account for "the laws of thought" (Boole).

---

[1]Work on this paper has been supported by the grant No. 17-15645S of the Czech Science Foundation. I am grateful to Vladimír Svoboda and Georg Brun for valuable criticism of previous versions of the manuscript.

Since then, logic has become quite entangled with mathematics, and even many logicians working in departments of philosophy do quite a lot of maths. Therefore, nowadays, it might seem that logic indeed is a science, namely that it is part of mathematics. In this paper I will defend the thesis that though logic is a science (or something close to science), this is not because it would be a part of mathematics. I think that logic is no more a part of mathematics than, say, physics is – that the massive ways in which it has come to rest on mathematical methods does not make it collapse into mathematics any more than it does physics. What I am going to argue is that logic is much more like a natural science than like mathematics: that it accounts for an empirical domain using complicated mathematical models[2].

I am convinced that the subject matter of logic is, roughly, human reasoning – not in the sense of an inner, psychological process, but as an ongoing human social activity[3], also known as argumentation. This activity is based on our human language and especially on that its part that can be called "logical". It is the form of this part of language where our argumentative practices have got sedimented. Hence, I am convinced, it is this language and the ways competent speakers use it that is the ultimate subject matter of logic. True, most of the studies of logic are legitimately carried out through various kinds of analyses of mathematical models of natural language and of reasoning, but the results of the analyses are *logical* results (rather than purely *mathematical* ones) to the extent – and only to the extent – to which it is the theory of our actual reasoning.

This is not to conceal that there is the important difference between physics and logic in that the latter, unlike the former, is *normative*. The normativity of logic comes in two varieties. One is that what logic accounts for are *norms* of our language and of our reasoning. This, *prima facie*, provides for a basic difference between logic and physics; but the difference may be smaller than it seems, if we accept that norms are just very complicated, feedback driven behavioral patterns structuring human societies (Peregrin, 2014). A more substantial difference consists in the fact that logic can influence its subject matter – that it may serve as a *norm* of reasoning in the sense that reasoners may take logic as telling them how they *should* reason. This is where the analogy between logic and a natural science like physics reaches its limits.

---

[2] See Shapiro (2001) for a similar view
[3] This sense of "reasoning" is epitomized by Laden (2012).

Logic as a (Natural) Science

2 Logic and reasoning

The general idea that logic is something like a theory of reasoning, in itself, is certainly in no way new or surprising. This plausible sounding claim, however, may easily lure us into oversimplified views of how exactly it accounts for reasoning. So before we go on, we should make it plain that such oversimplified pictures are not what we are aiming at.

The most straightforward understanding of logic as a theory of reasoning is the understanding of laws of logic as direct instructions for how to infer beliefs from beliefs; or, more generally, how to upgrade the set of one's beliefs so that it be maximally feasible. According to such interpretation, the law of the form

(1) $\quad p_1, \ldots, p_n \vdash p$

would amount to the instruction

(1*) if you hold all of the beliefs $p_1, \ldots, p_n$, hold also the belief p!

Thus, for example, the law of *modus ponens*,

(MP) $p, p \to q \vdash q$

would tell us that if we have the beliefs of the form p and $p \to q$, we should also have the belief q.

This view of the laws of logic has been shattered most notably by Harman (1986), and it is definitely not something we would like to suggest. The discussion emerging in the wake of Harman's book has indicated that the relation between logic and reasoning is quite complex[4]; and that there is no easy way to modify the above picture to make it feasible.

One reason why it is not reasonable to interpret (1) as (1*) is that the beliefs $p_1, \ldots, p_n$ as well as the belief p can be odd and we can hold $p_1, \ldots, p_n$ for various odd reasons; and it would hardly be viable if logic were to tell us to hold another odd belief. From this viewpoint, it might seem that a better interpretation of (1) might be[5]

---

[4] See, e.g., van Benthem (2008); Milne (2009); Field (2009); Dutilh Novaes (2015); or Steinberger (2019).

[5] Broome (1999) speaks about a "non-detaching relation" here: for in contrast to the previous case, which amounted to *if you believe p, you ought to believe q*, where p was "detachable" in the sense that if you do believe p, you have the obligation to believe q, this case amounts to *you ought to see to it that if you believe p, you believe q*, and given you believe p there is not an unambiguous obligation to adopt the belief q – if q is too weird, you can do justice to your obligation by giving up the belief p. The former case is also often called a *narrow scope* requirement, while the latter is dubbed a *wide scope* requirement (according to the scope taken by the *ought* – see, e.g., Way, 2010).

(1**) if you hold all of the beliefs $p_1, \ldots, p_n$, either hold also p, or abandon at least one of $p_1, \ldots, p_n$!

But the problem is that even if we hold perfectly reasonable beliefs $p_1, \ldots, p_n$, it would still not be reasonable for us to hold every p that follows from them. Any sentence has an unlimited number of trivial consequences (p_1, for example, has the consequences $p_1 \wedge p_1$, $p_1 \wedge p_1 \wedge p_1$, ...) and the obligation to hold all of them would lead to what Harman aptly described as "cluttering one's mind with trivialities" (Harman, 1986, p. 12)

Another option we might want to try is:

(1***) do not hold all of $p_1, \ldots, p_n$ together with $\neg p$!

But this already presupposes that we identify negation independently of principles of this kind, which is a problematic presupposition – any feasible definition of negation will already have to contain the clause that $p_1, \ldots, p_n \vdash \neg p$ renders p incompatible with $p_1, \ldots, p_n$ or something equivalent to it. (A proof-theoretic definition will probably contain $p, \neg p \vdash q$, which entails this via cut; a semantic definition will have to include something as $||p|| \cap ||\neg p|| = \emptyset$ which has a similar effect.)

Thus, interpreting the slogan *logic is the theory of reasoning* in some such straightforward way is problematic; and we repeat that this is *not* what we are after. Hence the question that comes into the fore is how exactly to understand it. In which sense is logic a theory of reasoning? Let us look at some recent attempts at an answer to it.

3 Williamson on logic

In a recent paper, Timothy Williamson (2017) expresses a view of logic *prima facie* very similar to that advertised above. He speaks about the "abductive methodology" that should govern our choice of logical theory, as well as governing the choice of any other scientific theory: we should judge logical theories, just as we judge theories in other sciences, "with respect to how well they fit the evidence, of course, but also with respect to virtues such as strength, simplicity, elegance, and unifying power". Hjortland (2017) summarizes this kind of attitude in the slogan "logic is not special": "Its theories are continuous with science; its method continuous with scientific method. Logic isn't a priori, nor are its truths analytic truths. Logical theories are revisable, and if they are revised, they are revised on the same grounds as scientific theories."

This seems to fit very well with the approach sketched above. Yes, log-

Logic as a (Natural) Science

ical theories are theories that are to systemize and explain certain bodies of evidence with certain aims, just as physical theories are to systemize the evidence we gain from perceiving, measuring, and experimenting in the physical world with the aim of discovering natural laws and helping us utilize them to foster our goals. But surprisingly, after claiming that the conceptual apparatus of logic should be instrumental to the formulation of "the most fruitful questions and their answers" and that it should be evaluated "with respect to how well they fit the evidence", Williamson does not go on to clearly articulate exactly what kinds of questions logic should answer and what kinds of evidence it is to fit.

What he puts forward is that the crucial theses logic is to accept or reject are universal generalizations of logical laws, such as "for every p, p or not-p". Such theses, Williamson argues, are not "meta-linguistics"; they are not about any language, rather they are about the world. In this way, Williamson comes close to the view of B. Russell (1919), who famously claimed that "logic is concerned with the real world just as truly as zoology, though with its more abstract and general features" (p. 169). (Wittgenstein, 1956, §I.8, ridiculed this construal of logic as the construal of "logic is a kind of ultra-physics, the description of the 'logical structure' of the world, which we perceive through a kind of ultra-experience".)

A crucial question, it would seem to me, is what the "for every p" in the above thesis quantifies over. The answer cannot be "everything", for the entities are subjected to logical operations (negation, disjunction), and hence they must something of the kind of sentences or propositions. If they are propositions, we can only access them via (such or another) sentences, so we can – and indeed must – keep using sentences as their proxies. Thus even if logical theses are not directly meta-linguistics, they are bound to be about entities intimately related to language (being subject to negation, disjunction etc.) and accessible only via linguistic entities (unlike other things of the world, which can be pointed at and investigated without the mediation of language).

And indeed, if we see logical laws as thus inherently related to language, we can see that there is no straightforward answer to the question about the domain of quantification of the "p" in Williamson's example. Of course, we have artificial languages of logic for which (or for the propositions expressed by their sentences) it holds. But we also have those in which it does not hold; and very probably it does not hold exceptionlessly for any natural language (i.e., for the propositions expressed by its statements). It would seem probable that every natural language contains sentences generally taken, by its speakers, as neither true nor false .... The principle can also be read as a methodological directive, as saying, roughly: doing logic, restrict your attention only to sentences that have one of the two truth values!

In any case, here it is where the view of logic put forward here differs essentially from Williamson's. Logic, by my lights, is about reasoning and argumentation; and reasoning (in the sense relevant for logic) and argumentation are things we do with (a) language[6]. It is reasoning and argumentation that is the basis for determining which kinds of logical theories are worth pursuing; just like any other theory, logic should be answerable to the relevant evidence and to the aims which guide our efforts to master its subject matter. And the evidence relevant for logic resides in the ways people actually reason and argue.

Hence, though like Williamson I do think there are important parallels between logic and physics (parallels ignored or denied by those who think that logic *is* special, that, e.g., it is a matter of an *a priori* analysis), I disagree with him concerning how exactly they are parallel. Williamson argues that logic should not "tailor its basic theoretical terms to fit whatever pre-theoretic prejudices and stereotypes may happen to be associated with the word 'logic', any more than physics should tailor its basic theoretical terms to fit whatever pre-theoretic prejudices and stereotypes may happen to be associated with the word 'physics' "; and though I certainly agree that logicians should not pay attention to "prejudices and stereotypes", I think this does not mean that they should not pay attention to how people really reason, to which arguments they hold for correct – for these form the *evidence* logic works with.

Why should we pay attention to the actual ways of reasoning of fallible people – should we not be interested only in the ways in which it is *correct* to reason, independently of whether anybody actually does reason in this way? Is what people actually do not irrelevant to what they should do – and is not paying attention to what they do just buying their "prejudices"[7]?

Consider purely practical (in contrast to theoretical or discursive) reasoning. I want to find out what to do to fulfill my needs. Here it might be purely the study of the world that is able to show us how to fulfill the needs

---

[6] As Mercier and Sperber (2017) put it: "Unlike verbal arithmetic, which uses words to pursue its own business according to its own rules, argumentation is not logical business borrowing verbal tools; it fits seamlessly in the fabric of ordinary verbal exchanges. In no way does it depart from usual expressive and interpretive linguistic practices". (p. 172)

[7] Nowadays we have access to many studies of the fallacies human reasoners tend to make (from Wason, 1968, or Kahneman & Tversky, 1996, to a host of their more recent followers).

Logic as a (Natural) Science

most effectively, especially what kind of reasoning is most efficient. However, most of our reasoning is theoretical and discursive, and to do this it employs meaningful symbols. And to be able to say what is the best way to operate with certain *symbols* we need to know what they *mean*. And to learn what the symbols mean we need to pay attention to how their users actually use them, including how they reason with them.

We carry out all our theoretical reasoning in terms of concepts, mostly the concepts bestowed on us by our mother languages. (We might have developed some amended versions of the languages with slightly better – clearer, less ambiguous ... – concepts, but the heritage of our normal language is certain to still play a vital role.) Hence, if we are to be given any advice on how to reason, it will have to be an advice concerning how to reason within the framework of our natural language.

Many of the important theses our reasoning focuses on will probably be of the shape *If A then B*. If somebody wants to advise us how to deal with these more proficiently, she or he will have to know what exactly they mean – in particular, what exactly *if ... then ...* means. And to learn this, she or he will have to gather evidence concerning how we actually deal with the expressions – and in this particular case also what kinds of arguments including *if ... then ...* we hold for correct, which for incorrect and which are perhaps indeterminate.

4   G. Russell on the nature of logical laws

Another recent paper treating logic as a theory of reasoning is due to G. Russell (2015). Considering the nature of logical laws and rejecting two *prima facie* plausible accounts of logical laws (namely, the account that logical laws are analytic truths which implicitly define the meanings of logical constants and the account that they are simply some very central nodes of our overall web of belief), the author comes to the conclusion that "logic isn't basic, reasoning is". Russell's approach then comes closer to the one put forward here than Williamson, however, it does not bring it to its consequences.

Russell describes a four stage pilgrimage of a student of logic into the secrets of the subject. In the first stage, the student enters as a complete layman, her "beliefs about logic are rather inchoate". In the second stage, she discovers classical logic and starts to consider it as *the* logical tool, "she enthusiastically accepts both the general theory of truth-functional logic, and the more specific claim that the law of excluded middle is a logical truth." The third stage is marked with the recognition that some rules of classical logic, especially the rule of excluded middle (hereafter EM), may not be valid after all, and she "eventually comes to agree with [her professor] that classical logic is wrong, and she should adopt the three-valued logic instead." In the fourth stage she is confronted with paraconsistent logic, which subverts still more laws of classical logic, but leads to a logic that is "intolerably weak", so she starts to reconsider her rejection of EM and "she steps back to classical logic, holding ultimately, that its theoretical virtues and power outweigh those of the alternatives" and she "regains her belief that the law of excluded middle is a logical truth".

Thus, the way in which Russell sees reasoning as setting the agenda of logic is that it is the *de facto* reasoning that determines what is logically valid and what is not, and thus it underlies the content of logical laws. This, I think, is very true, but it seems to me that neither Russell's story, nor the morals she draws from it, are as clear and as explicative as they should be. What are the main lacunas I see in Russell's exposition?

Firstly, Russell wants to expose the nature of logical laws, but in the end she does *not* tell us, explicitly, what the laws are. Instead she gives us her story, from which she draws certain morals, but surprisingly no explicit moral with respect to the nature of the laws. Her story is instructive and can help us to gain insight into the nature of logic; but an explicit conclusion seems to be lacking.

Secondly, I think that to reach such an explicit conclusion would call for a more careful scrutiny of the nature of the concept of "validity" (or that of "logical truth", which the author seems to use interchangeably with "validity"), which features prominently in the story (indeed it is the "coming-of-age story" in the course of which the hero moves from naive views of what is "valid" to more mature views) and which, I am afraid, harbors certain dangerous ambiguities.

Anyway, the fact that Russell puts the concept of validity in the center of her picture deserves a special attention. On the one hand, it is in no way surprising: that the question which logical principles are valid (or which are – the true? – laws of logic) is central to logic appears to go without saying. On the other hand, however, if we view logic as a theory of the *de facto* enterprise of reasoning, this is no longer so obvious – should not logic, in this case, concentrate on some regularities of this very process?

I think that the solution to this quandary lies in the recognition of the fact that the *de facto* reasoning is an (essentially) *rule-governed* enterprise. We play our "games of giving and asking or reasons" (Brandom, 1994) according to certain rules, just as we play football according to its rules. (And here we should not imagine so much an "official" football all the rules of which are explicitly canonized; but rather a "yard game", with most of the rules relatively clear but not explicitly spelled out.) And central to the account for the game is capturing the rules that constitute its framework. Thus, the way we account for the enterprise of reasoning is that we try to capture its rules, by means of what we call logical validities or the rules of logic.

Hence the picture that looms before us is the following: As a matter of fact, we humans engage in ongoing reasoning, in the never-ending game of "giving and asking for reasons". The rules are part of the game, just like those of football are part of football – though most of them are not necessarily quite explicitly written down, they are implicit in that the players respect them and are prone to ostracize their violators. The most basic aim of a theory of reasoning, hence, is to make the rules fully explicit: to present the model that captures them as adequately as possible and as reasonable. And it captures them in the form of schemata which we tend to call *valid* and which we tend to see as *logical laws*.

Given this understanding of the enterprise of logic, it is, first and foremost, necessary to be quite clear about what *validity* amounts to. The question is whether this concept, as standardly construed within logic, can be taken as the very concept that can play the role of the principal explicandum of logic understood as a theory of *de facto* reasoning.

5 What kind of validity?

Rules of our "giving and asking for reasons" get captured, within a formal language, as certain patterns: schematic arguments or directly as schematic statements. Thus, *modus ponens* gets captured as

(MP) $p, p \to q \vdash q$

whereas *excluded middle* gets captured as

(EM) $p \lor \neg p$

where "p" and "q" are contentless placeholders. Such a schematic argument or a schematic statements is then called valid if all its instances are correct arguments or true statements.

One concept of validity, hence, is straightforward – the concept internal to a logical system. Thus we know, for example, in this sense that EM is valid in classical logic and invalid in intuitionistic logic[8]. Can we use this relative concept of validity to arrive at an absolute one? Certainly, it is enough to raise a particular logical system to the rank of *the true* logic. Then the validity within this system becomes validity *simpliciter*. However, it is not clear which decisive arguments could support such a claim of a logical system. Of course, it might be that somebody sees a logical system as embodying some mental operations crucial for logical reasoning; or it might be that one sees it as capturing some logical relationships that "in fact" interconnect Platonic ideas or propositions; but, as any evidence for either the former or the latter view could only be very indirect and susceptible to differing interpretations, it would be hard to see it as indisputably supportive.

Alternatively, we can try to go for an absolute concept of validity from the beginning. We might say that the letter "p" in "$p \vee \neg p$" stand for some "real" sentences or propositions, and hence that the law says that the disjunction of a sentence or a proposition with its own negation is always true. What kind of entity would this formulation refer to? They might be linguistic entities (sentences of a natural language) or entities related to the linguistic ones (propositions expressed by such sentences), but as we have already noted, there is no way of getting hold of the latter directly, without the mediation of the former, and hence it would seem reasonable to focus our attention on the linguistic entities, sentences.

Construed thus, the validity of EM comes down, within our linguistic milieu, to the claim that the disjunction of any English (declarative) sentence with its own negation is (necessarily?) true. This, to be sure, presupposes that we know what the disjunction of two English sentences is, and what the negation of an English sentence is; and while for the former we can say that it is the complex sentence which arises out of connecting the sentences with

---

[8] Of course, it is slightly less straightforward to delimit this concept generally. We might try: EM is valid in a logical system iff the formula "$p \vee \neg p$" is a theorem/tautology of the system. However, this would obviously not work, for the specific signs a system employs are arbitrary, and "$\vee$" and "$\neg$" need not express disjunction and negation, respectively. Hence what we need would be: EM is valid in a logical system iff the formula "$p \vee \neg p$" (or, for that matter, "$p \oplus \circledast p$"), where "$\vee$" ("$\oplus$") is a disjunction of the system and "$\neg$" ("$\circledast$") is its negation, is a theorem/tautology of the system. And then we would have to define what it takes to be a *disjunction* and a *negation* of a logical system, which, in general, is far from straightforward. (Notice that someone might want to say that the validity of EM is one of the *defining* features of – "genuine" – negation. Notice that we might have systems where more constants aspire to being a disjunction or a negation, etc.) But let us leave all these difficulties aside and assume that this relative concept of validity is straightforward.

Logic as a (Natural) Science

the connective "or", the latter is more problematic. (For example, is "The king of France is not bald" the negation of "The king of France is bald"?) Hence in this case the validity of EM is not quite non-negotiable. In the case of many sentences of natural language we have significant leeway over the determination of both where to place the boundary of true sentences and which sentences are to count as instances of "$p \lor \neg p$".

One might try to argue that the problem results from our decision to focus on sentences rather than on propositions. Only some sentences express propositions, the argument might go, and it is perhaps only those not expressing them that constitute the invalid instances of "$p \lor \neg p$". However, there is no simple way to draw a dividing line between those sentences that do, and those that do not express propositions;[9] and as a result, this argument appears to be a sleight of hand: if we dismiss any counterexamples on account of their "not expressing propositions" without being able to formulate any *independent* criterion of when a sentence does express a proposition, we are effectively turning EM into an irrefutable claim which therefore ceases to be interesting.

Now validity within a formal system (which is a matter of the definition of the system) and validity within natural language (which is a matter of a general claim which can be tested empirically) are two very different notions. They can coincide – if we fine-tune the formal system so that its validities capture precisely (or at least approximate reasonably) the validities of the natural language. However, such an absolute coincidence is unlikely – for the replication of all the twists and turns of any natural language would force our logic to be overly complicated and heterogeneous, while logic should be, as any other model, something simple and perspicuous.

Thus, a logical analysis, the subsuming of natural language cases under the umbrellas of formulas of a formal language (which are then considered the "logical forms" of the natural language sentences) usually involves plenty of mutual adaptation. It is not only that logic is formed to comply with the language, but that the language, conversely, must be as if "pressed into a suitable conceptual mold" – and sometimes even directly regimented

---

[9] Proposals to this effect often contend that only sentences that are in some sense *complete* express propositions. (Thus, as Frege, 1918, p. 76, for example, puts it, the sentence "this tree is covered with green leaves" is not complete, because it does not specify the "the time-indication".) However, then the question is whether *any* sentence of natural language is sufficiently complete to express a unique proposition. The sentence "this tree is covered with green leaves" is certainly *not* complete until the tree in question is uniquely specified; but it is also not quite clear what the boundaries are for a tree being "covered" by green leaves (in contrast to merely "having" green leaves) etc.

– to allow for a relatively simple coverage by logical forms. This process is very similar to the process of reflective equilibrium which is argued to yield ethical rules.[10]

Note also that the kind of "harmony" between a natural language and logic which would substantiate us in saying that a logic captures a language, is not something the existence of which could be proved or ascertained once and for all. The relation between a formal language and a natural one is like that between a formal model of an empirical phenomenon and the phenomenon itself: we can carry out various kinds of "measurements" to examine whether the former captures the latter, but no amount of such measurement is able to establish that it is *absolutely* adequate. Moreover, what we call *adequate* is a matter of the purpose for which we do the modeling.

Unfortunately, many freshmen in logic, and also at least some of its veterans, seem to have the intuition not only that the two notions of validity can be calibrated and maintained in the required harmony, but that this harmony is somehow intrinsic. They take it, for example, that the "→" of classical (or, for that matter, another) logic is *intrinsically* tied to the English "if ... then ...", just because they are two incarnations of a supernatural *implication* ("→" being definite and perfect, and "if ... then ..." being loose and elusive). Probably nobody would really want to defend this explicitly, as its oddness is readily seen, but nevertheless it seems to lurk in the background of a lot of thoughts about logic. This is documented by the "unbearable lightness" with which many textbooks on logic move back and forth between sentences with "if ... then ..." and corresponding formulas with "→", as if they were naturally the same.

6 Two kinds of languages

Considerations of the previous paragraph come down to the essential importance of the distinction between two kinds of languages. Every human (perhaps with some negligible exceptions) speaks at least one language – a language she has not invented, but which was passed to her by her elders. The expressions of the language are meaningful, and they are passed from one generation to another as such. (One may take part in a process in which the meanings of some expressions get gradually modified, but the bulk of

---

[10] See Peregrin and Svoboda (2016, 2017). In fact though it was Rawls (1971) who coined the term, the idea behind it was exploited already earlied, by Goodman (1983), and it was precisely in connection with logic.

Logic as a (Natural) Science

the meanings are simply not up for grabs for any given individual; they are, as it were, prescribed to her.[11]) It is in terms of these meanings that she reasons, thinks and understands her world.

Can one *create* a language? Certainly one can create something that shares enough features with natural languages to be called *language* – be it something like Esperanto or something like the language of Peano arithmetic. But can one create a language with semantics as comparably rich and nuanced as that of natural language? This is much more dubious. In general, we may endow expressions of an artificial language with meanings either by linking them to expressions of a natural language and thus borrowing their meanings (as in the case of Esperanto) or by trying to craft the meanings from scratch. When we go for the former option, the question is whether what we gain is a truly new self-standing language, or rather a mere simulacrum parasitic on an existing one or existing ones; if we go for the latter, the question is whether we are able to create something that will be really usable as a language, or, indeed, be of any use at all.

There is no problem in sitting down and devising a "language" by positing a vocabulary, some grammatical rules and some rules of semantics – be it a kind of "model-theoretic" semantics imitating the relation of designation or rather a "proof-theoretic" one related to the use-theory of meaning. The question is whether such a "language" would be of any use. We know that it may be useful if it is close enough to a part of our natural language that it can be used as its simplified (and/or more precise, more perspicuous ...) "model", or even as its more rigorous "proxy". This, for example, is the case for an artificial language of Peano arithmetic, which is so closely related to our pre-theoretical discourse in natural language that its sentences can be taken as precisely expressing what we had expressed imprecisely before. This may also be the case for various languages of pure logic (which however, are usually not genuine languages but rather mere language forms).

It seems clear that when we ask whether EM holds, we cannot mean whether it holds in an arbitrary artificial language. We know we can easily construct an artificial language (or 'language'?) in which it holds, as well as we can equally easily construct another one in which it does not hold. Moreover, there is no saying which kind of language is a "genuine" language of logic – we know that both languages supporting EM, such as classical logic,

---

[11] And let me stress that this is no mysticism of the kind of "the genuine language is shrouded in mystery". Natural language is simply so complex, and so complexly interwoven with all the things we do, that it is hardly possible to recreate it, in all its complexity, in "laboratory conditions".

and languages rejecting it (such as intuitionistic or three-valued logic) have firm places in the enterprise of logic. And we have also seem that looking at EM as summarizing truths of natural language heralds many problems.

As a result, we have artificial languages, where logical laws hold (or do not hold) in clear, unambiguous and often provable ways, which, however do not seem to have any universal bindingness for us; and we have natural languages which do bind us (in that they are our universal medium of grasping the world), in which logical laws hold (or do not hold) only fuzzily and usually not exceptionlessly. And as the question about the validity of a law like EM does not seem to fit with the context of either of the kinds of languages, it might seem it must go with something that is "beyond" the languages, to something which is both precise and binding enough to make the question straightforward. It might be a language of thought with which each of us is born, or a system of propositions within a Platonist heaven towards which the minds of each of us work their way. But the trouble with ideas like these is that they are just *ad hoc* stipulations fashioned to make our philosophical life easier. There is no way to investigate such an "absolute" language in an independent and intersubjective way.

Instead of this, I think that the *locus* of validity of logical laws can be seen as the interaction of the natural and the artificial languages. A rule of a formal language becomes a logical law insofar as the formal language becomes our standard "prism" through which we see our natural language. Of course, the choice of such a prism is far from arbitrary – far from every artificial language is usefully employable as such a prism. However, there is still leeway: as we know, there are logical accounts of our linguistic intercourse, of our reasoning and of our cognitive life based on classical logic as well as those based on non-classical logic which rejects EM.

7 Logic as a science

It is time to return to the elucidation of the claim that logic, in some, important aspects, is like a natural science. The picture we have sketched up to now is that the formal languages that have been ubiquitous in logic during the last hundred-odd years are like the mathematical models that have become ubiquitous in physics or other natural sciences.

Thus, we can agree that logic – in this sense – is not special, as Hjortland stresses, and that it follows the kind of abductive methodology urged by Williamson. Logic tries to cope with the evidence and yield theories that are able to systematize it in the simplest, most effective and most usable way. However, as most of the reasoning that is its subject matter rides the vehicle of language, it is basically about language and about rules that govern some of our language games. (Also, it may sometimes suggest how to improve on them, or how to play them using some artificially created linguistic items instead of those offered by ordinary language. However, such improvements and gadgets are usually only local and do not wholly disentangle us from the framework of our ordinary language.) I am not sure whether this means that logic is, in Williamson's term, "meta-linguistic", but it does mean that it is largely concerned with language.[12] (Though insofar as language is part of the real world, logic *is* concerned with the world, albeit merely with a part of it.)

The situation in physics is such that there are some data, data which usually result from measuring some parameters of some natural objects, events of phenomena, and these data are used to build the model. The advantages of the model are that it is simpler than the phenomenon itself, it is explicitly and exactly delimited, and it is susceptible to mathematical treatment. (Of course it is crucial that it is simpler in just the right way, that it disregards those features of the phenomenon which are not important from the current viewpoint and retains those that are.) Then we can use mathematical methods to learn something new about the model, and project these results back on the phenomenon which it was a model of.

The thesis now is that the formal languages of logic can be seen precisely as models in this sense, as models of "natural" reasoning or argumentation and of its "natural" vehicles, natural languages. What remains to be clarified is what exactly counts as the data which are both the point of departure of building the models and its checkpoints.

Priest (2016, p. 355) writes: "In the criterion of adequacy to the data, what counts as the data? It is clear enough what provides the data in the case of an empirical science: observation and experiment. What plays this role in logic? The answer, I take it, is our intuitions about the validity or

---

[12]Quine (1960, p. 273) stresses that this does not render logic as a matter of linguistic in any deep sense: "Most truths of elementary logic contain extralogical terms; thus 'If all Greeks are men and all men are mortal ...'. The main truths of physics, in contrast, contain terms of physics only. Thus whereas we can expound physics in its full generality without semantic ascent, we can expound logic in a general way only by talking of forms of sentences. The generality wanted in physics can be got by quantifying over non-linguistic objects, while the dimension of generality wanted for logic runs crosswise to what can be got by such quantification. It is a difference in shape of field and not in content; the above syllogism about the Greeks need owe its truth no more peculiarly to language than other sentences do."

otherwise of vernacular inferences." This is almost precisely what I want to propose. I say *almost*, because I do not think it is good to use the term "intuition". I would say that what counts as data is "the validity or otherwise of vernacular inferences", i.e., which inferences in natural languages are taken and treated as correct. This can be researched empiricially, and indeed it can be researched by "observation and experiment" (which reemphasizes the proximity of logical research and the research in natural sciences). We can observe which inferences are used and accepted in real arguments and we can set up experiments to find out which such inferences would be accepted or considered correct.

Let me repeat that this rapprochement of logic and natural sciences does not do away with the pending *dis*similarities between logic and natural sciences mentioned above: especially with the fact that unlike theories in natural sciences, logical theories do not only *capture* its subject matter, but rather *interact* with it. They can be used to *correct* human reasoners thereby having the feedback on their subject matter. This does render the whole enterprise different from that of natural sciences. Hence in contrast to the "non-exceptionalist" program I maintain that there *is* a discontinuity between the method of logic and those of the sciences; though I agree that the extent to which they are continuous is important and interesting.

8   The nature of logical laws

If we take natural language at face value, then almost none of the laws articulated by common logical systems, were we to apply them as strictly as possible, would hold for it. (We have seen this for the case of the EM, but counterexamples have been reported with respect to almost any logical law, including MP.) The reason for this is that the inferential properties of "logical" expressions of natural language are generally far more complex (and also less determinate) than those of the logical constants that we normally use to regiment them.

What does this show us regarding the nature of logical laws? Put briefly, I think that what we take to be a logical law is a rule of a formal language that we find indispensable for systematizing certain basic rules of natural language. Thus MP is a logical law as it is hard to imagine a formal language capable of usefully formalizing the whole of natural language which would lack an implication governed by this law. It follows that a logical law is a law not so much in the sense of a natural law, i.e., of a discovered natural regularity, but rather in the sense of a linguistic rule, which is first present, in a tentative form, in our natural language, and then fortified and raised to a true law when canonized by our logical theory.

There is one more misconception we should avoid when taking the laws of logic as rules of reasoning. We have already rejected the view that the rule tell us directly which beliefs to hold. However, here we must reject even the more general view that the laws are any kind strategic instructions telling us how to manage the system of our beliefs to keep it in good shape. From the viewpoint advocated here, this is not true, for the laws are not strategic, but rather *constitutive* rules, rules constitutive of (the meanings of) logical constants. Hence, rather than telling us how to think or how to manage our beliefs, they produce, as it were, "material" with which to think, of which to compose some complex beliefs.

Take MP: It is properly so called only insofar as → is an *implication*. (Note that were it a conjunction, the rule would still be valid – but *such* a rule is certainly not what we call MP.) But it is hard to imagine how to characterize implication, as contrasted with other kinds of operators, without including MP. (Someone might object that we can find a logical system in which implication does not obey MP, or does not obey it unexceptionally. The reply is that if there is an implication not obeying MP and if it is still to be a specific kind of operator distinguishable from other operators, then there must be some *other* rules – or at least features of its logical behavior – that characterize it. And I do not see anything that might be generally acceptable; so I think that it is necessary to stick to MP and to conclude that if an operator is called implication and does not obey MP, then it is so called only by courtesy.[13])

Here we may instructively refer to the well worn comparison of the laws of logic with the rules of chess. Laws like MP or EM are much more similar to the rules constitutive of chess (the rules delimiting the permitted kinds of moves) than to the rules that would instruct us how to play chess so as to improve our chance of winning. And just as we can see the constitutive rules of chess as "producing" individual pieces, like pawns, rooks, bishops etc., which are what we can then rally to lace into the opponent, so we can see the laws of logic as "producing" logical constants, like negations, conjunction, implication, etc., which we can then use to engage in our human kind of "logical", "propositional" or "rational" thinking, especially reasoning.

---

[13] In natural language, on the other hand, we can assume only an approximate match – we cannot assume that natural language expressions will exactly fit with the artificial categories.

9 Conclusion

I think that logic is secondary to reasoning: with a certain oversimplification we can say that logic is to reasoning as physics is to the swarming of the natural world. Logic offers theories of reasoning, analogously to physics offering theories of behavior of spatiotemporal things. And just as physics also does, logic accounts for its subject matter in terms of idealized models, with the consequence that the laws it formulates apply immediately and unexceptionally to the models, but only in a mediated way to the original subject matter.

And once we distinguish between the natural languages, which are the natural vehicles of our reasoning, and artificial languages, which act as the models, we can depict the parallel even much more concretely. The role of the data (which, in case of physical theories result from various kinds of measurements of the world) is played by the fact concerning the inferences that are taken and treated for correct by the speakers of the natural languages (which, again, can be measured, though in this case by methods of sociology, which are less exact and reliable than those used by physics).

There are, of course, important differences between logic and physics. A crucial one is that what logic accounts for are rules, which can be themselves influenced by logical theories. Thus – like in ethics and unlike in physics – there is a feedback in which the theories may make us not only recategorize the data, but modulate – be it only slightly – the stream of the data, by changing the behavior of the subjects who produce them. In this sense the reflective equilibrium yielding our logical laws not only homes in on the most effective conceptual framework for our accounts, but also plays an active role in what is to be accounted for.

Also like in ethics and unlike in physics, the laws of logic advise us what to do, in particular how to reason. And here we must be careful not to mistake logical laws, which mostly delimit the space in which reasoning may take place, and which constitute the equipment needed to do so, for rules that advise us how to reason effectively and fruitfully. From this viewpoint, the laws of logic can mostly be seen as constitutive of the meanings of logical constants – *viz.* as analytic truths implicitly defining them.

It is, however, necessary to keep in mind that what is in play are always *two* kinds of languages: there is the messy, but conceptually binding natural language and there are the exact, but unbinding formal languages. It is only when we achieve the required harmony between them, making one of the latter ones a prism through which we see the former one (the fine-tuning of the harmony being achieved in the process of the reflective equilibrium), that the two very different entities interconnect to yield something that is both exact and binding – and what we tend to call logical laws.

References

Brandom, R. (1994). *Making it Explicit: Reasoning, Representing, and Discursive Commitment*. Cambridge (Mass.): Harvard University Press.

Broome, J. (1999). Normative requirements. *Ratio, 12*(4), 398–419.

Dutilh Novaes, C. (2015). A dialogical, multi-agent account of the normativity of logic. *Dialectica, 69*(4), 587–609.

Field, H. (2009). What is the normative role of logic? *Proceedings of the Aristotelian Society, Supplementary Volume, 83*(1), 251–268.

Frege, G. (1918). Der Gedanke. *Beiträge zur Philosophie des deutschen Idealismus, 2*, 58–77. (English translation: The Thought, *Mind, 65*(259), 1956, 289–311.)

Goodman, N. (1983). *Fact, Fiction, and Forecast*. Cambridge (Mass.): Harvard University Press.

Harman, G. (1986). *Change in View (Principles of Reasoning)*. Cambridge (Mass.): MIT Press.

Hjortland, O. T. (2017). Anti-exceptionalism about logic. *Philosophical Studies, 174*(3), 631–658.

Kahneman, D., & Tversky, A. (1996). On the reality of cognitive illusions. *Psychological Review, 103*(3), 582–591.

Laden, A. S. (2012). *Reasoning: A social picture*. Oxford: Oxford University Press.

Mercier, H., & Sperber, D. (2017). *The Enigma of Reason*. Cambridge (Mass.): Harvard University Press.

Milne, P. (2009). What is the normative role of logic? *Proceedings of the Aristotelian Society, Supplementary Volume, 83*, 269–298.

Peregrin, J. (2014). Rules as the impetus of cultural evolution. *Topoi, 33*(2), 531–545.

Peregrin, J., & Svoboda, V. (2016). Logical formalization and the formation of logic(s). *Logique et Analyse, 59*(233), 55–80.

Peregrin, J., & Svoboda, V. (2017). *Reflective Equilibrium and the Principles of Logical Analysis: Understanding the Laws of Logic*. New York: Routledge.

Priest, G. (2016). Logical disputes and the apriori. *Logique et Analyse*, *59*(236), 347–366.

Quine, W. V. O. (1960). *Word and object*. Cambridge (Mass.): MIT Press.

Rawls, J. (1971). *A theory of justice*. Cambridge (Mass.): Harvard University Press.

Russell, B. (1919). *Introduction to Mathematical Philosophy*. London: Allen & Unwin.

Russell, G. (2015). The justification of the basic laws of logic. *Journal of Philosophical Logic*, *44*(6), 793–803.

Shapiro, S. (2001). Modeling and normativity: How much revisionism can we tolerate? *Agora*, *20*(1), 159–173.

Steinberger, F. (2019). Consequence and normative guidance. *Philosophy and Phenomenological Research*, *98*(2), 306–328.

van Benthem, J. (2008). Logic and reasoning: Do the facts matter? *Studia Logica*, *88*(1), 67–84.

Wason, P. C. (1968). Reasoning about a rule. *The Quarterly Journal of Experimental Psychology*, *20*(3), 273–281.

Way, J. (2010). Defending the wide-scope approach to instrumental reason. *Philosophical Studies*, *147*(2), 213.

Williamson, T. (2017). Semantic paradoxes and abductive methodology. In B. Armour-Garb (Ed.), *Reflections on the Liar* (pp. 325–346). Oxford: Oxford University Press.

Wittgenstein, L. (1956). *Bemerkungen über die Grundlagen der Mathematik*. Oxford: Blackwell. (English translation: *Remarks on the Foundations of Mathematics*, Blackwell, Oxford, 1956.)

Jaroslav Peregrin
Czech Academy of Sciences, Institute of Philosophy
The Czech Republic
E-mail: peregrin@flu.cas.cz

Non-Constructive Procedural Theory of Propositional Problems and the Equivalence of Solutions

Ivo Pezlar[1]

Abstract: We approach the topic of solution equivalence of propositional problems from the perspective of non-constructive procedural theory of problems based on Transparent Intensional Logic (TIL). The answer we put forward is that two solutions are equivalent if and only if they have equivalent solution concepts. Solution concepts can be understood as a generalization of the notion of proof objects from the Curry-Howard isomorphism.

Keywords: Transparent Intensional Logic, procedural semantics, logic of problems, procedural isomorphism

1 Introduction and motivation

There can be many different solutions to a single problem. For example, consider the problem whether $\sqrt{2}$ is an irrational number or not: we have geometric solutions, algebraic solutions, constructive solutions, etc.[2] But are all these solutions really different? In some cases, it seems easy to decide, but in other cases, it is not so obvious. Interestingly, these difficulties are not exclusive to relatively advanced mathematical problems and they appear at the level of the simplest logical problems as well.[3] Consider, e.g., the following two solutions to the problem $A \to ((A \to B) \to B)$ carried out in a natural deduction system for propositional logic:

---

[1] Work on this paper was supported by Grant No. 17-18344Y from the Czech Science Foundation, GA ČR.

[2] See, e.g., (Harris, 1971).

[3] In the case of propositional problems, solutions can be understood simply as proofs and problems as propositions. However, since this relation does not generally hold for all problems and solutions (e.g., it seems reasonable to say that 12 is a solution to the problem expressed by $5 + 7$, it makes less sense to say that 12 is a proof of $5 + 7$), we keep this terminology.

Ivo Pezlar

$$\text{Solution A } \cfrac{\cfrac{\cfrac{A \to B \quad A}{B} \to E}{(A \to B) \to B} \to I}{A \to ((A \to B) \to B)} \to I \qquad \text{Solution B } \cfrac{\cfrac{\cfrac{\cfrac{A \to B \quad A}{B} \to E}{A \to B} \to I \quad A}{B} \to E}{\cfrac{(A \to B) \to B}{A \to ((A \to B) \to B)} \to I} \to I$$

Are these two solutions equivalent? At first glance we would probably say yes. After all, they start with the same premises and end with the same conclusion. But on the other hand, they are also clearly syntactically distinct: one has more steps than the other.

The received view would tell us that these two solutions are equivalent because they can be converted into each other. More specifically, the solution **B** can be reduced to **A** by removing all the 'unnecessary detours' (so-called *normalization* procedure). In this case, it is the first application of the implication introduction rule immediately followed by the application of the corresponding implication elimination rule. If we cut it away from **B**, we get **A**.[4] This approach is also corroborated by the propositions as types principle.[5] If we decorate the above derivations with the corresponding λ-terms (also called *proof objects*), we obtain:

$$\text{Solution A1 } \cfrac{\cfrac{\cfrac{x : A \to B \quad y : A}{x(y) : B} \to E}{\lambda x.x(y) : (A \to B) \to B} \to I}{\lambda y.\lambda x.x(y) : A \to ((A \to B) \to B)} \to I$$

$$\text{Solution B1 } \cfrac{\cfrac{\cfrac{\cfrac{x : A \to B \quad y : A}{x(y) : B} \to E}{\lambda y.x(y) : A \to B} \to I \quad y : A}{(\lambda y.x(y))(y) : B} \to E}{\cfrac{\lambda x.(\lambda y.x(y))(y) : (A \to B) \to B}{\lambda y.\lambda x.(\lambda y.x(y))(y) : A \to ((A \to B) \to B)} \to I} \to I$$

The concluding term $\lambda y.\lambda x.(\lambda y.x(y))(y)$ of the second solution can be rewritten into the concluding term $\lambda y.\lambda x.x(y)$ of the first solution. More specifically, $\lambda y.\lambda x.(\lambda y.x(y))(y)$ is reducible to $\lambda y.\lambda x.x(y)$ via β-reduction. Hence, these two terms are considered to be equivalent and, consequently, so are the corresponding solutions.[6]

---

[4] See, e.g., (Prawitz, 2006).

[5] Also known as the Curry-Howard isomorphism or correspondence (Curry & Feys, 1958; Howard, 1980).

[6] The equivalence of λ-terms has been thoroughly studied, most notably by Church himself (Anderson, 1998; Church, 1954, 1993).

Non-Constructive Procedural Theory of Propositional Problems

Our approach to solution equivalence based on Transparent Intensional Logic will also utilize λ-calculus, however, it will allow for partial functions and have a much more semantic flavor. For us, two solutions will be considered equivalent if and only if they have equivalent *solution concepts*. Intuitively, solution concepts can be understood as reified methods or procedures for solving problems. From a technical standpoint, they can be regarded as abstract generalizations of proof objects that need not be effective, i.e., constructive in the intuitionistic sense. This will enable us to analyze even incorrect solutions.[7]

2 TIL: the basics

Transparent Intensional Logic (TIL)[8] is a many-sorted type theory with hyperintensional semantics initially devised for natural language analysis. Similarly to Montague semantics, it utilizes λ-calculus but makes room for partial functions. The central notion of TIL is a construction, which can be understood as a reified abstract algorithm.[9] Constructions are assigned to linguistic expressions as meanings and encode procedures for determining their denotations. For example, propositions are understood as procedures for computing truth values (see also, e.g., Jespersen, 2017; Moschovakis, 2006; Muskens, 2005).[10]

TIL typically relies on six fundamental kinds of constructions, however, we will use a generalized presentation requiring only four constructions:

$$constructions := x^\alpha \mid [C^\alpha C_1^{\beta_1} \ldots C_m^{\beta_m}] \mid [\lambda x_1^{\beta_1} \ldots x_m^{\beta_m} C^\alpha] \mid {}^n X^\alpha$$

where x is a variable, C_i is any construction, X is either a construction or a non-construction (i.e., an object that is not a construction, e.g., truth value, individual, number, etc.), and α, β_i are type metavariables.[11] The first three constructions are called *variable*, *composition*, and *closure* and they roughly

---

[7]We will leave the notion of incorrect solutions intentionally vague to not limit ourselves to some specific conception of solution correctness.

[8]See (Tichý, 1988), and also (Duží, Jespersen, & Materna, 2010; Raclavský, Kuchyňka, & Pezlar, 2015).

[9]Not to confuse with intuitionistic constructions, which are essentially effective proofs.

[10]We will not go through the specifics of TIL-based analysis of natural language here, since it has been well covered in many other places already (see, e.g., Duží et al., 2010; Raclavský et al., 2015).

[11]The notation X^α means that X is an object of type α if X is a non-construction, otherwise it means that X is a construction typed to v-construct an object of type α. We will omit the type letters 'α', 'β', ... to simplify the notation. For a proper specification of constructions, see

correspond to variable, function application, and function abstraction from λ-calculus.[12] The last one, called *n-execution* (alternatively, *n*-stage execution or multiple execution), is a distinctive feature of TIL: given the procedural nature of constructions, this construction allows us to either execute them (possibly more than once) to find out what they construct, i.e., what is their output (if $n > 1$), or leave them as they are, i.e., don't execute them (if $n = 0$). What particular objects constructions construct may depend on a *valuation v*, i.e., an assignment of values to free variables. If that is the case, we say that they *v-construct* those objects.[13] For example, let us have some variable construction x and some valuation v that assigns to it some object X, then we can say that the construction x v-constructs X. If a construction C v-constructs nothing at all, we will say that it is a *v-improper* construction. Otherwise, we say that C is a *v-proper* construction. If we have two constructions C_1 and C_2 that v-construct the same object X, or they are both v-improper, we will say that C_1 and C_2 are *v-congruent* constructions, denoted as $C_1 \cong C_2$. If they are v-congruent for all valuations v, we will say that C_1 and C_2 are *equivalent*, denoted as $C_1 = C_2$.

For example, the 0-execution construction $^0 12$ yields the number 12. Intuitively, '$^0 12$' can be read as 'do not execute 12, just refer to it'. And we don't want to execute 12, because it is not a construction but a number, i.e., a non-construction.[14] By definition, the 1-execution construction $^1 12$ constructs nothing, i.e., it is an improper construction (we cannot execute non-constructions, only constructions). Analogously, the composition construction $[^0+\ ^0 5\ ^0 7]$ constructs 12 and so does, e.g., the 1-execution $^1[^0+\ ^0 5\ ^0 7]$. Hence, we can say that they are congruent, and even equivalent constructions. On the other hand, the 0-execution $^0[^0+\ ^0 5\ ^0 7]$ constructs $[^0+\ ^0 5\ ^0 7]$, but, e.g., 2-execution $^2[^0+\ ^0 5\ ^0 7]$ is again an improper construction (2-execution is essentially a two stage execution: the first stage gives us 12 and the second stage consists of its 1-execution $^1 12$. However,

---

(Tichý, 1988), (Duží et al., 2010) or (Raclavský et al., 2015). For a specification of *n*-execution, see (Pezlar, 2018).

[12] Although in many situations λ-terms and constructions can behave in similar fashion (e.g., both are open to α-, β-, η-conversions – this will be important later in Section 3.1), there are non-trivial conceptual as well as technical distinctions. Most importantly, constructions are not terms but abstract objects that can be executed to yield some other objects.

[13] To be more precise, constructions always construct with respect to some valuation v. In some cases, however, valuation does not affect on the overall result of the construction. If that is the case, we will simply speak of *constructing* instead of *v-constructing*.

[14] Note that 0-execution can play roughly the same role as do constants in impure/applied λ-calculus.

as we already know, this is an improper construction, hence the whole construction is improper).

To simplify the notation, we will denote 0-execution by **boldface** font, with the exception of standard connectives and operators such as $+$, $=$, $\forall$, $\rightarrow$, etc. that will be kept in normal font with 0-execution implicitly assumed. Also we will use infix notation whenever anticipated. For example, we will write $[\mathbf{5 + 7}]$ instead of $[^0+\ ^05\ ^07]$ and $[A \supset B]$ instead of $[^0\supset A\ B]$.

All objects including constructions receive a type. If α and $\beta_1, \ldots, \beta_m$ are types, then $(\alpha\beta_1\ldots\beta_m)$ is also a type. Specifically, a type of function from the elements of type $\beta_1, \ldots, \beta_m$ to the elements of type α. For example, the construction $[^0+\ ^05\ ^07]$ has type $*_1$ (so-called 1st-order construction),[15] while $^0[^0+\ ^05\ ^07]$ has type $*_2$ (2nd-order construction). On the other hand, non-constructions $+$, 5, and 7 have types $(\nu\nu\nu)$, ν and ν, respectively, where ν is the type of natural numbers.

2.1 Solution assignment

In (Pezlar, 2017) it was shown that the previous TIL-based non-constructive procedural approach to analysis of logical problems (see, e.g., Materna, 2004, 2008) is too coarse-grained because it renders every solution to every logical problem as equivalent. The same paper tries to alleviate this issue by introducing a new notion called *solution constructor* that helps us to track the process of solution construction in a similar way as we would do with the Curry-Howard correspondence (free variables corresponding to assumptions, etc.).[16] The solution constructor can be used to analyze correct solutions, however, here we generalize it further into so-called *solution assignment* that can analyze any solution, both correct and incorrect. Our rationale is the following: dealing with incorrect solutions—or more precisely, with solutions that are believed to be correct, but later are shown to be incorrect—is a natural aspect of problem-solving (similarly as is, e.g., debugging in programming) and by limiting our framework to correct solutions only we are unnecessarily restricting its expressive power.

---

[15]More specifically, the construction $[^0+\ ^05\ ^07]$ has type $*_1$ as its lowest possible order of type. Due to TIL's cumulative hierarchy of types, it is also of type $*_2, *_3, \ldots$, etc. To put it simply, any construction of type $*_n$ is also a construction of type $*_{n+1}$. For more about the ramified hierarchy of types in TIL, see (Tichý, 1988).

[16]Conceptually, the main change is that instead of λ-terms and propositions (as their types) we work with higher- and lower-order constructions, hence proof objects are considered as higher-order objects than the things they are proving. Technically, there are some slight differences induced by TIL idiosyncrasies. More on this below.

Definition 1 (Solution assignment) *A solution assignment is a couple $c :: C$ where C is a propositional construction (a problem to be solved) and c is any construction v-constructing a construction of the same type as C (if any). A solution assignment $c :: C$ is said to be* satisfied *if c v-constructs C. Otherwise, we say that it is* unsatisfied.

Note that if c in $c :: C$ is an improper construction, then the whole solution assignment is unsatisfied (follows from Def. 1 and the definition of improper constructions). Also note that a solution assignment $c :: C$ can be satisfied even though C itself is an improper construction. Conveniently, this gives us a way to analyze situations when an agent tries to solve nonsensical, or more precisely, truth-valueless problems such as 'the largest prime number ends with the number 7', '3 divided by 0 equals 2', etc. Furthermore, solution assignments allow even 'bad' solutions such as solving C by considering some trivial construction of C (e.g., the 0-execution of C). Informally, this sort of degenerate solutions can be likened to the colloquial 'it is solved because I said so' method. Shortly put, with solution assignments we just check if c yields C, not whether c is also a good solution of C. Hence, $c :: C$ should not be read as 'c solves C' or 'c is a solution to C' but rather as 'c is an assumed solution of C', 'c is considered as a solution to C' or 'c solves C, allegedly' and similarly with the option that it might be false.[17] The separation of the good solutions from the bad ones will be dealt with at the level of solutions concepts.[18]

Definition 2 (Solution concept) *Let us have a solution assignment $c :: C$, then we will say that c is a* solution concept *of the problem C. If c solves C,[19] then we will say that c is a* suitable *solution concept. Otherwise, we will say that c is an* unsuitable *solution concept.*

At first glance, the notion of solution concept might seem enigmatic. There is, however, a simple idea behind it: it is a procedure, not necessarily

---

[17] We could restrict this condition and require that for every solution assignment $c :: C$ holds that c solves C. For example, Constructive Type Theory (see Martin-Löf, 1984) can be seen as a system where this restriction holds, i.e., every solution must be effective.

[18] At the LOGICA 2018 conference, prof. Duží posed the following question (my paraphrase): Cannot we avoid the necessity for the solution assignment, and thus for the use of higher-order constructions, simply by invoking the notion of refinement of constructions as defined in Duží et al. (2010), Chapter 5? Unfortunately, the answer is no. Although we could refine the logical connectives used, it would not help us to track the actual solution construction, which is our main objective.

[19] Or maybe more precisely, if c can be recognized as a solution of C (see, e.g., Wittgenstein, 1922, 6.2321 or Wittgenstein, 1978, II-42).

Non-Constructive Procedural Theory of Propositional Problems

effective, that represents solution construction – recall the dictum proofs-as-programs from the Curry-Howard isomorphism. To put it differently, we can think of solution concepts as abstract generalizations of proof objects from general proof theory. In contrast to them, solution concepts can be applied even to non-logical problems (although in this paper, we focus only on the logical ones) and can be defective. From this point of view, proof objects can be understood as a special case of solution concepts that are always effective.[20] Compare this with, e.g., Constructive Type Theory (CTT), where every proof object must be effective, i.e., terminate in a finite amount of steps, which is not the case for solution concepts. On the other hand, similarly to proof objects in CTT, solution concepts are proper objects of TIL: they receive types, we can use them as arguments for higher-order functions, etc. (this will be important later in Section 3.1, where we discuss equivalence of solution concepts).

Note that the adoption of solution concepts helps us to explain why can we 'comprehend' even incorrect solutions, i.e., follow them, find mistakes in them, etc. From our perspective, they are not meaningless or nonsensical. They have an idea—some solution concept—behind them, it just happens to be an ineffective one that does not do the required job, i.e., a solution concept that does not solve the problem at hand.

As we already mentioned, solution assignments are used to emulate in TIL the solution tracking behaviour of the Curry-Howard isomorphism. We demonstrate this by formulating rules for implication, conjunction, disjunction, and negation capable of recording the solution forming process. We start with the implication introduction rule. Suppose that A and B represent any propositional constructions (i.e., constructions v-constructing the truth values $true$ or $false$ of type o), x and y are variables ranging over the objects of type $*_n$ (i.e., variables for constructions), and let $\supset$ be implication of type (ooo). Next, we establish premises for our implication introduction rule. Suppose that we have some solution assignment $x :: A$ such that x is of type $*_{n+1}$ and that it v-constructs A of type $*_n$. Further, assume we derive from it another solution assignment $y :: B$ such that y is of type $*_{n+1}$ and that it v-constructs B of type $*_n$ (thus, both are satisfied solution assignments). So we have:

---

[20]Historically, solution concepts can be loosely regarded as an explication of Wittgenstein's approach: 'The idea that proof creates a new concept might be also roughly put as follows: a proof is not its foundations plus the rules of inference, but a *new* building [= our solution concepts] ...' (Wittgenstein, 1978, II-41).

$$(x :: A)$$
$$\vdots$$
$$y :: B$$

Now, we introduce the function ii of type $(*_n(*_n*_n))$ that will imitate the behaviour of the implication introduction rule. More concretely, ii is a higher-order function that takes a function—representing an inference from A to B—v-constructed by $[\lambda x\, y]$ and outputs a construction representing a conclusion of this derivation, otherwise it is undefined. If we put it together, we get the construction $[\text{ii}\ [\lambda x\, y]]$ that v-constructs $[A \supset B]$. The corresponding rule $\supset$**I** then looks as follows:[21]

$$\frac{\begin{array}{c}(x :: A)\\ \vdots \\ y :: B\end{array}}{[\text{ii}\ \lambda x\, y] :: [A \supset B]}\ \supset\text{I}$$

Now, we move to the implication elimination rule. We start again by establishing the premises. Suppose we have two solution assignments $x ::$ $[A \supset B]$ and $y :: A$ such that x and y are of type $*_{n+1}$ and that they v-construct $[A \supset B]$ and A, both of types $*_n$, respectively (i.e., they are satisfied solution assignments). Thus we have:

$$x :: [A \supset B] \qquad y :: A$$

Next, we introduce a higher-order function that will mimic the behaviour the implication elimination rule. Concretely, we introduce the function ie of type $(*_n\ *_n\ *_n)$ that takes two constructions of the form $[A \supset B]$ and A as arguments and outputs another construction of the form B, otherwise it is undefined. The resulting construction $[\text{ie}\ x\, y]$ then v-constructs B. As the corresponding rule $\supset$**E** we get:[22]

$$\frac{x :: [A \supset B] \qquad y :: A}{[\text{ie}\ x\, y] :: B}\ \supset\text{E}$$

---

[21] We omit the outer brackets of the closure construction when in composition with ii or other constant to get a cleaner notation.

[22] Why do we need to work with constructions of the form $[\text{ie}\ x\, y]$ instead of the more traditional form $[x\, y]$ as known from the Curry-Howard correspondence? From a TIL perspective, $[x\, y]$ is a composition construction, which means that its first component (x in this case) has to v-construct a function, otherwise it would be improper. Since in our rule $\supset$**E** the variable x v-constructs $[A \supset B]$, which is not a function, we have to introduce the auxiliary function ie to circumvent this issue. Similarly in the case of ii.

Non-Constructive Procedural Theory of Propositional Problems

To recapitulate, the ii and ie are higher-order functions that capture the behaviour of the corresponding rules. More specifically, ii is a function of type $(*_n(*_n*_n))$ that takes a function (representing the inference of B from A) as an argument and returns a new construction of the form $[A \supset B]$. Analogously, ie is a function of type $(*_n *_n *_n)$ that takes two arguments of the form $[A \supset B]$ and A and returns a new construction of the form B.

The rules for conjunction, disjunction and negation will be presented more succinctly since they follow the same ideas as the rules above and share the general form with their counterparts from the Curry-Howard isomorphism. We start with conjunction:

$$\frac{x :: A \quad y :: B}{[\textbf{ci}\ x\ y] :: [A \wedge B]} \wedge\text{I} \qquad \frac{z :: [A \wedge B]}{[\textbf{pr}_l\ z] :: A} \wedge\text{E}_l \qquad \frac{z :: [A \wedge B]}{[\textbf{pr}_r\ z] :: B} \wedge\text{E}_r$$

The **ci** constructs the 'conjunction introduction rule' function in a similar way as did **ii** previously for implication introduction rule. The components $\textbf{pr}_l$ and $\textbf{pr}_r$ construct the familiar left and right projection functions. Types of all these functions can be easily inferred: **ci** has type $*_{n+1}$ and constructs a higher-order function of type $(*_n *_n *_n)$ and $\textbf{pr}_l$ and $\textbf{pr}_r$ are also of type $*_{n+1}$ and construct higher-order functions of type $(*_n *_n)$. Next, we introduce the rules for disjunction:

$$\frac{x :: A}{[\textbf{j}\ x] :: [A \vee B]} \vee\text{I}_l \qquad \frac{y :: B}{[\textbf{k}\ y] :: [A \vee B]} \vee\text{I}_r$$

$$\frac{c :: [A \vee B] \quad \begin{array}{c}(x :: A)\\ \vdots \\ d :: C\end{array} \quad \begin{array}{c}(y :: B)\\ \vdots \\ e :: C\end{array}}{[\textbf{de}\ c\ \lambda x\ d\ \lambda y\ e] :: C} \vee\text{E}$$

The **j** and **k** are constants that tell us from which higher-order construction was the disjunction constructed, the **de** then constructs the 'generalized disjunction elimination rule' function of type $(*_n *_n\ (*_n*_n)(*_n*_n))$. Finally, the rules for negation are analogous to the rules for implication:

$$\frac{\begin{array}{c}(x :: A)\\ \vdots \\ y :: \bot\end{array}}{[\textbf{ii}\ \lambda x\ y] :: [\neg A]} \neg\text{I} \qquad \frac{y :: [\neg A] \quad x :: A}{[\textbf{ie}\ y\ x] :: B} \neg\text{E}$$

205

The construction $[\neg A]$ is equivalent to $[A \supset \bot]$ and $\bot$ is to be understood as a construction of a proposition that is always false.

With inference rules ready, we can now define solutions in a more precise way:

Definition 3 (Solution) *A solution (or a solution tree) is a finite sequence of solution assignments $\langle c_1 :: C_1, \ldots, c_n :: C_n, c :: C \rangle$ each of which solution assignment is either an axiom, or an assumption, or follows from the preceding solution assignments in the sequence by some inference rule of the system. The last solution assignment will be also called conclusion.*

A solution tree can be also written vertically as $\dfrac{c_1 :: C_1 \quad \ldots \quad c_n :: C_n}{c :: C}$.

For example, the solution **A** can be analyzed in TIL in the following way:

Solution A'
$$\dfrac{\dfrac{\dfrac{x :: [A \supset B] \quad y :: A}{[\text{ie } x\, y] :: B}\supset\text{E}}{[\text{ii } \lambda x\, [\text{ie } x\, y]] :: [[A \supset B] \supset B]}\supset\text{I}}{[\text{ii } \lambda y\, [\text{ii } \lambda x\, [\text{ie } x\, y]]] :: [A \supset [[A \supset B] \supset B]]}\supset\text{I}$$

Note that each solution assignment is satisfied and every corresponding solution concept is suitable.

To incorporate even incorrect solutions, we have to appropriately modify the definition of solution trees. Specifically, by allowing to add to the sequence solution assignments that are neither axioms, assumptions, or follow from the inference rules. As an example of an incorrect solution, or more specifically, of an incorrect solution step, consider the following:

$$\dfrac{\begin{array}{c} A \\ \vdots \\ B \end{array}}{[A \wedge B]}$$

where the conjunction $[A \wedge B]$ is incorrectly inferred instead of the implication $[A \supset B]$. In TIL, we can analyze it as follows:

$$\dfrac{\begin{array}{c} x :: A \\ \vdots \\ y :: B \end{array}}{[\text{ci } \lambda x\, y] :: [A \wedge B]}$$

Non-Constructive Procedural Theory of Propositional Problems

Notice that the concluding solution concept is unsuitable and the solution assignment is unsatisfied: the function constructed by **ci** expects two arguments but receives only one. In other words, the solution concept [**ci** $\lambda x\, y$] is an improper construction. However, it does not mean that [**ci** $\lambda x\, y$] as such becomes nonsensical in TIL. It is still a construction in its own right, even though improper, i.e., we can use it as an object of predication. For example, we can introduce a higher-order unary function Improper of type $(o*_n)$ for checking properness of constructions. If we feed it the construction [**ci** $\lambda x\, y$], i.e., form the composition [**Improper** ^{0}[**ci** $\lambda x\, y$]], we get the value $true$.

3 Solution equivalence

When do we generally consider two solutions as equivalent? Arguably, we would say that two solutions are equivalent when they follow the same general method. As mentioned above, we code this method via higher-order TIL constructions called solution concepts. Thus the issue of solution equivalence is reduced to the question of solution concept equivalence.

Definition 4 (Solution equivalence) *Solutions S_1 and S_2 are equivalent if and only if the solution concepts appearing in their conclusions are equivalent.*

Observe that the initial question 'When are two solutions equivalent?' is thus transformed into 'When are two solution concepts used by two different solution trees equivalent?' which will be the focus of the following section.

3.1 Fine-tuning the solution equivalence threshold

Let us return to the solutions **A** and **B**. Applying the approach discussed above, we learn that their respective solution concepts are:

[**ii** λy [**ii** λx [**ie** [**ii** λy [**ie** $x\, y$]] y]]]   and   [**ii** λy [**ii** λx [**ie** $x\, y$]]]

Solution concepts were designed to emulate proof objects based on λ-terms. Naturally, we can reuse the criteria for equivalence of λ-terms as well with appropriate changes where necessary. The standard conception is to regard λ-terms as equivalent if they are λ-convertible, i.e., α-, η- and β-convertible. We can carry over this general approach to TIL as well. However, increased caution is necessary due to the fact that β-conversions and η-conversions are generally not equivalent transformations in TIL.[23] For these reasons we exploit here the variant of construction equivalence put forward by Duží and Jespersen (2012) called procedural isomorphism alternative ($\frac{3}{4}$) with the difference that we drop η-reduction.[24]

Definition 5 (Procedural isomorphism) *Let C and D be constructions. Then C and D are procedurally isomorphic if and only if either C and D are identical (i.e., the same construction) or there are constructions $C_1, \ldots, C_n$ ($n > 1$) such that $^0C = {}^0C_1$, $^0D = {}^0C_n$, and for each C_i, C_{i+1} ($1 \leq i < n$) it holds that C_i, C_{i+1} are either α-equivalent (i.e., they differ only by having different λ-bound variables) or β_r-equivalent (i.e., equivalent via restricted β-conversion by name).*

We prefer this variant of procedural isomorphism to the later one proposed by Duží and Jespersen (2015) and dubbed alternative (A1"). Even though it also omits η-conversion, it relies on β-conversion by value which is not a best fit for us – we are not really interested in what the solution concepts construct (i.e., what are their 'values'), but rather in the solution concepts themselves (i.e., in their 'names'). For that reason we choose β_r-conversion, which is essentially just a tool for a formal simplification of constructions (see Duží & Kosterec, 2017) guaranteeing that all the resulting transformations of constructions will be equivalent even in the presence of partial functions and potentially improper constructions and that is all we need for analyzing propositional logical proofs.

We write the fact that two constructions C and D are α-, β_r-equivalent as $C =_\alpha D$ and $C =_{\beta_r} D$, respectively.

Definition 6 (Equivalence of solution concepts) *Solution concepts c_1 and c_2 are equivalent if and only if they are procedurally isomorphic. We denote this as $c_1 \equiv c_2$.*

It follows that when we say that two solutions are equivalent it means that their respective solution concepts are procedurally isomorphic. Note that if c is a suitable solution concept for C and $c_1 \equiv c_2$, then c_2 is also a suitable solution concept for C. Also note that if c_1 and c_2 are procedurally isomorphic ($c_1 \equiv c_2$), then c_1 and c_2 are congruent ($c_1 \cong c_2$), but not vice versa.

---

[23] Or any other logic of partial functions (see, e.g., Moggi, 1988).

[24] In theory, other alternatives could be chosen as well. Duží (2017) provides a great survey of possible variants. Our approach essentially corresponds to Duží's variant **C6**, modulo meaning postulates.

Let us now return to our initial question: are the solutions **A** and **B** equivalent? From the perspective of TIL, they are equivalent because their corresponding solution concepts are procedurally isomorphic.

4 Conclusion

In this paper we expanded upon the non-constructive procedural approach to analysis of problems proposed in our previous work (Pezlar, 2017) by addressing the issue of solution equivalence. In short, we take two solutions as equivalent if and only if their solution concepts are equivalent, i.e., procedurally isomorphic. Solution concepts can be understood as a generalization of the notion of proof objects from the Curry-Howard isomorphism. The main difference is that solution concepts need not be effective, which allows us to analyze even cases involving incorrect solutions. Future work lies mainly in extending our framework towards predicate logic, incorporating mathematical problems, and finally investigating the possibility of expanding our approach towards empirical problems as well.

References

Anderson, C. A. (1998). Alonzo Church's contributions to philosophy and intensional logic. *Bulletin of Symbolic Logic, 4*(2), 129–171.

Church, A. (1954). Intensional isomorphism and identity of belief. *Philosophical Studies, 5*(5), 65–73.

Church, A. (1993). A revised formulation of the logic of sense and denotation. Alternative (1). *Nous, 27*(2), 141–157.

Curry, H. B., & Feys, R. (1958). *Combinatory Logic* (Vol. 1). North-Holland Publishing Company.

Duží, M. (2017). If structured propositions are logical procedures then how are procedures individuated? *Synthese, 196*(4), 1249–1283.

Duží, M., & Jespersen, B. (2012). Transparent quantification into hyperintensional contexts de re. *Logique et Analyse, 55*(220), 513–554.

Duží, M., & Jespersen, B. (2015). Transparent quantification into hyperintensional objectual attitudes. *Synthese, 192*(3), 635–677.

Duží, M., Jespersen, B., & Materna, P. (2010). *Procedural Semantics for Hyperintensional Logic: Foundations and Applications of Transparent Intensional Logic*. Dordrecht: Springer.

Duží, M., & Kosterec, M. (2017). A valid rule of β-conversion for the logic of partial functions. *Organon F, 24*(1), 10–36.

Harris, V. C. (1971). On proofs of the irrationality of the square root of 2. *The Mathematics Teacher, 64*(1), 19–21.

Howard, W. A. (1980). The formulae-as-types notion of construction. In H. B. Curry, J. R. Hindley, & J. P. Seldin (Eds.), *To H. B. Curry: Essays on Combinatory Logic, Lambda Calculus, and Formalism.* Lodon: Academic Press.

Jespersen, B. (2017). Anatomy of a proposition. *Synthese, 196*(4), 1285–1324.

Martin-Löf, P. (1984). *Intuitionistic Type Theory.* Bibliopolis.

Materna, P. (2004). *Conceptual Systems.* Logos.

Materna, P. (2008). The notion of problem, intuitionism and partiality. *Logic and Logical Philosophy, 17*(4), 287–303.

Moggi, E. (1988). *The Partial Lambda-Calculus* (Unpublished doctoral dissertation). Faculty of Mathematics and Informatics, University of Edinburgh.

Moschovakis, Y. N. (2006). A logical calculus of meaning and synonymy. *Linguistics and Philosophy, 29*(1), 27–89.

Muskens, R. (2005). Sense and the computation of reference. *Linguistics and Philosophy, 28*(4), 473–504.

Pezlar, I. (2017). Algorithmic theories of problems. A constructive and a non-constructive approach. *Logic and Logical Philosophy, 26*(4), 473–508.

Pezlar, I. (2018). On two notions of computation in transparent intensional logic. *Axiomathes, 29*(2), 189–205.

Prawitz, D. (2006). *Natural Deduction: A Proof-theoretical Study.* Dover Publications, Incorporated.

Raclavský, J., Kuchyňka, P., & Pezlar, I. (2015). *Transparentní intenzionální logika jako characteristica universalis a calculus ratiocinator.* Brno: Masaryk University Press (Munipress).

Tichý, P. (1988). *The Foundations of Frege's Logic.* Berlin: de Gruyter.

Wittgenstein, L. (1922). *Tractatus Logico-Philosophicus.* London: Kegan Paul, Trench, Trubner & Co., Ltd.

Wittgenstein, L. (1978). *Remarks on the Foundations of Mathematics* (Revised ed.). Oxford: Basil Blackwell.

Ivo Pezlar
Czech Academy of Sciences, Institute of Philosophy
The Czech Republic
E-mail: pezlar@flu.cas.cz

A First-Order Sequent Calculus for Logical Inferentialists and Expressivists

SHUHEI SHIMAMURA[1]

Abstract: I present a sequent calculus that extends a nonmonotonic reflexive consequence relation as defined over an atomic first-order language without variables to one defined over a logically complex first-order language. The extension preserves reflexivity, is conservative (therefore nonmonotonic) and supraintuitionistic, and is conducted in a way that lets us codify, within the logically extended object language, important features of the base thus extended. In other words, the logical operators in this calculus play what Brandom (2008) calls expressive roles. Expressivist logical systems have already been proposed for propositional logics (see Hlobil, 2016, and Kaplan, 2018) but not for first-order logics. An advantage of this calculus over standard first-order calculi (e.g., those in Gentzen, 1935/1964) is that universally quantified variables behave as they should even in the presence of arbitrary nonlogical axioms. I claim that because of this robust well-behavedness of variables, this calculus also provides logical inferentialists with a way to understand the meanings of variables in terms of the roles those variables play in a wide range of inferences that is not limited to purely logical ones (e.g, mathematical inferences).

Keywords: nonmonotonic logic, first-order logic, sequent calculus, logical inferentialism, logical expressivism

1 Introduction: Two philosophical motivations

1.1 Logical inferentialism

Variables seem to play an essential role in various phases of our linguistic practices. This is most evident in mathematical practices, where we explic-

---

[1] This paper is the product of a joint work with the research group of Bob Brandom. The technical results reported here are mine. For valuable discussions and comments, I thank Bob Brandom, Ulf Hlobil, Dan Kaplan, Rea Golan, Stephen Mackereth, Mansooreh Kimiagari, Adrian Anhalt-Gutierrez, Yao Fan, and the audience of Logica 2018 conference and several other conferences and workshops at which earlier versions of this paper were presented.

itly employ variables for particular aims that do not seem accomplishable by other means. For example, when we want to prove that for all right-angled triangles, the square of the hypotenuse is equal to the sum of the squares of the other two sides, it seems necessary to reason with an arbitrary right-angled triangle, say, t. Similarly, when we want to specify an equation or a function, it seems inevitable for us to use variables, "x", "y", "z", etc., to talk about those arbitrary relata that are related in a particular manner within the equation or function at issue. Here, variables seem to play a distinctive role—they function as though referring to arbitrary objects.

However, it is not only in mathematics that we need to talk or think about arbitrary objects; such occasions are prevalent in our ordinary linguistic practice. Consider, for instance, how we explain the meaning of "match" to children. We say something like "if you strike a match, it lights." We are talking about neither this or that match, but rather an arbitrary match (otherwise, this explanation would be of little use). Thus, although the original sentence does not explicitly contain a variable, it seems to say something that is more explicitly said by using a variable: "For any x, if x is a match and you strike x, then x lights."

Variables seem to let us talk about objects without specifying them. This seemingly distinctive function of variables, however, perplexed Bertrand Russell—one of the first philosophers to notice the great potential use of variables for the analysis of natural-language sentences (see, e.g., Russell, 1905). This is because he also clung to the view that the meaning of an expression is specified in terms of what it represents or refers to. What then does a variable, say "x," mean? This question confronts Russell with a formidable dilemma (see, e.g., Russell, 1994, p. xxxv). Apparently, no particular object counts as the proper referent of "x." If the meaning of "x" is specified by its referent, it follows that "x" has no specifiable meaning. Or one may bite the bullet here and claim that there exist arbitrary objects along with usual particular ones, and "x" refers to one of them. This horn, however, immediately invites many tough questions, such as where and how such arbitrary objects exist, how they can ever be distinguished from each other despite their arbitrariness, and so on.[2] Thus, both horns appear difficult to grasp. Let us call this *Russell's dilemma of the meaning of variables.*

Several attempts have been made to solve Russell's dilemma by seeking an account of the meanings of variables while maintaining his representa-

---

[2]Frege (1979, p. 160), for instance, expresses his doubts about the notion of arbitrary objects.

A First-Order Sequent Calculus for Logical Inferentialists/Expressivists

tionalist assumption that meanings are explained in terms of references (e.g., Fine, 1985, 2007). However, one of the two main aims of this paper is to propose a different way out of this dilemma. It seems to me quite natural to regard variables as (parts of) logical operators. According to some, the meanings of logical operators may be explained more naturally by looking at the rules governing their proper inferential use than by looking for things that they might refer to. A common example is the conjunction. It is notoriously difficult to seek the referent of the conjunction. Nevertheless, we all seem to know what it means. What do we explain, then, when we explain the meaning of the conjunction? A natural answer seems to be the rule governing its proper use—in particular, the rule governing what a conjunctive is properly inferred from and what is properly inferred from it. This line of thought is sometimes called *logical inferentialism*.[3] What I pursue in this paper is a logical inferentialist approach to Russell's dilemma of the meanings of variables. That is, I explain the meanings of variables by looking for a set of rules governing their proper inferential use, rather than for what they might refer to.

One may wonder, though, if we already have such rules, because due to Gentzen (1935/1964) and his successors, there are suitable proof systems for various first-order logics in which such rules are conveniently isolated for the variable-involving logical operators.

Somewhat surprisingly, however, it is difficult to find a proof system in the literature that can fulfill the current aim. It seems that in order to evaluate whether given rules for the universal quantifier—one of the major variable-involving logical operators—do justice to its intuitive meaning, it is an essential criterion that those rules guarantee the following biconditional: Γ implies $\forall x A \Leftrightarrow$ for any a: Γ implies $A[a/x]$, where x does not freely occur in Γ. Let us call this *the universal principle*. As far as I know, the universal rules of most proof systems guarantee this principle only within the limits of purely logical inferences—namely, only under the condition that they are free from proper axioms (I discuss why in the next section). As the examples mentioned at the beginning of this paper illustrate, however, variables seem to play an essential role not only in purely logical inferences, but also in mathematical or even more casual inferences. Thus, logical in-

---

[3] See, e.g., (Peregrin, 2014). The possibility of logical inferentialism, which is inspired by the famous remark of Gentzen (1935/1964, p. 295) on his natural deductions ("The introductions represent, as it were, the 'definitions' of the symbols concerned ..."), has been investigated by many both philosophically and technically. Peregrin (2014, pp. 3–6) offers a nice overview of this research tradition that covers most recent works.

ferentialists who wish to systematically explain the meanings of variables in these wider contexts[4] need a new proof system equipped with a set of rules that guarantees the well-behavedness of universals more robustly.

1.2 Logical expressivism

The second important aim of this paper is to submit a proof system that can also help advance the enterprise of *logical expressivism*, proposed and developed by Brandom (1994, 2000, 2008). Once we widen our focus from purely logical inferences to more casual ones, we may also be able to widen the scope of inferentialism: we may be able to explain not only the meanings of logical vocabulary but also those of nonlogical vocabulary in terms of the roles they play in the widened range of inferences.[5] If the meanings of logical and non-logical vocabulary are explained alike in terms of their inferential roles, it may be wondered how these two types of vocabulary can be distinguished in the first place. Logical expressivism is an inferentialist answer to this demarcation problem of logic.

Brandom (2009, p. 11) aspires to characterize logical vocabulary as "the organ of semantic self-consciousness"—the organ that is potentially available to anyone who can talk and that, once actualized, lets one talk about what one means when one talks. In (Brandom, 2008, pp. 52–54), this slogan is cashed out as two conditions for a piece of vocabulary to count as logical. First, the inferential roles of sentences involving that would-be logical operator can be mechanically determined on the basis of the inferential roles of sentences without it. Thus, logical vocabulary comes for free, as it were, for anyone who master nonlogical vocabulary. This condition, which Brandom calls *algorithmic elaboration*, is shown to be met by providing a proof system for a language containing the would-be logical operator that systematically determines the inferential roles of all the sentences involving that operator on the basis of those of atomic sentences. Second, the would-be logical operator must let us codify, without disturbing it, some aspect of the inferential role of the nonlogical vocabulary. A typical instance of this role, which Brandom calls *explicitation* or *expression*, can be played by the conditional. In a proof system in which the so-called deduction theorem and its converse hold (i.e., Γ implies $A \to B \Leftrightarrow \Gamma \cup \{A\}$ implies B), the conditional lets us codify, within the object language, the information that

---

[4]It may be worth stressing here that Gentzen (1935/1964) himself explicitly states his intention to use his proof systems for the analysis of inferences in mathematics (p. 288, p. 291).

[5]Strictly speaking, this is what Brandom (2000, p. 28) calls "weak inferentialism."

A First-Order Sequent Calculus for Logical Inferentialists/Expressivists

one thing (i.e., A) implies another (i.e., B) in a given context (i.e., Γ). In this way, the conditional lets us talk about implications, which, according to inferentialism, (partly) constitute the meanings of the sentences invovled. The nondisturbance proviso is also met if the proof system is conservative.

The ambition of logical expressivists as sketched above imposes several requirements on a proof system. First, since logical expressivists want to talk about the meanings of nonlogical (as well as logical) vocabulary (such as "match"), it needs to deal with nonlogical (as well as logical) inferences in terms of which those meanings are supposed to be understood. For instance, to understand the meaning of "match," it seems that one needs to understand, say, that "a is a match" and "a is struck" jointly implies, other things being equal, "a lights." As this example illustrates, inferences of this type, which are sometimes called "material inferences,"[6] often seem to be defeasible, and therefore nonmonotonic. Thus, a logical expressivist's proof system must be able to accommodate such a nonmonotonic material consequence relation as proper axioms, and to conservatively extend it to a logically complex one. Let us call this constraint *Nonmonotonicity*. Second, as illustrated above, the deduction theorem and its converse are unnegotiable for logical expressivists, because otherwise the conditional cannot play its expressive role to codify an implication. Similar biconditional constraints are imposed on the negation and other logical operators that are supposed to express different aspects of the underlying material consequence relation. Let us call this constraint *Expressivity*.[7]

There are already several nonmonotonic proof systems equipped with logical operators playing different expressive roles, such as expressing implication, incoherence, local monotonicity (see, e.g., the original work by Hlobil, 2016 for a supraintuitionistic system and its extension—with several improvements—to a supraclassical system proposed by Kaplan, 2018). One limitation of these systems, however, is that they are all propositional logics. From the logical expressivist viewpoint, this means that we can only talk about the meanings of entire sentences (e.g., "a is a match"), but we cannot yet talk about the meanings of their component expressions (e.g., "match"). For if we want to talk specifically about the meaning of, say, "match," it seems that we must say something like this: "For any x, if x is a match and x is struck, then x lights." Given the universal principle, the universally quantified variable "x" here codifies, roughly, that the pattern

---

[6] See, e.g., (Sellars, 1953) and (Brandom, 1994, 2000, 2008).

[7] For a more rigorous and generalized characterization of *Expressivity*, see (Kaplan, 2018, sec. 2).

of inference stated above holds irrespective of the individual term involved. This operator thus seems to let us talk purely about the inferential role of "is a match" (in connection with the other predicates such as "strike" and "lights") while bracketing that of "a".

So far, we have seen how the two philosophical ideas, logical inferentialism and logical expressivism, motivate us to pursue a proof system in which the universal principle robustly holds even in the presence of arbitrary proper axioms. In the next section, I explain why most of the currently available proof systems fail to satisfy this demand. In section 3, I introduce my alternative nonmonotonic proof system. Finally, in section 4, I prove that this system has several desirable properties, such as preserving reflexivity, being conservative, and guaranteeing the universal principle even in the presence of arbitrary proper axioms, along with other such biconditional properties demanded by *Expressivity*. In that section, I also show that the system is supraintuitionistic.

2 The problem

Most of the currently available proof systems fail to assure the universal principle in the presence of some proper axioms. Although logical inferentialism has usually been pursued within the framework of natural deductions, I focus on sequent calculi because they can make the problem at issue more straightforwardly visible. As long as the different frameworks of proof systems capture the same logic, however, problems shown to occur in one must also occur in the other. Furthermore, as will be seen below, the problem arises easily from a few common features shared by many different logics, including classical, intuitionistic, relevant, and modal logics. Thus, I claim, the scope of the problem is quite wide.

In sequent calculi of various logics, the inferential contributions from the universal quantifier are usually specified by the following pair of left and right rules[8]:

$$\frac{\Gamma, A[\tau/\xi] \mathrel{\vert\!\sim} \Theta}{\Gamma, \forall \xi A \mathrel{\vert\!\sim} \Theta} \text{L}\forall \qquad \frac{\Gamma \mathrel{\vert\!\sim} A[\zeta/\xi], \Theta}{\Gamma \mathrel{\vert\!\sim} \forall \xi A, \Theta} \text{R}\forall$$

where no ζ freely occurs in Γ, $\forall \xi A$, or Θ

---

[8]Note that below I use the snake turnstile (i.e., $\mathrel{\vert\!\sim}$) instead of the regular turnstile (i.e., $\vdash$) to indicate that the consequence relation at issue can be nonmonotonic.

A First-Order Sequent Calculus for Logical Inferentialists/Expressivists

The additional clause of the right universal rule is the so-called "eigenvariable condition," which is supposed to secure the "arbitrariness" of the variable substituted (i.e., "ζ") and thereby justify its substitution by the universally quantified variable (i.e., "ξ"). Notice that in the standard setting it is only via free occurrences of eigenvariables that we can introduce universal quantifiers on the right. How, then, can free eigenvariables be introduced in the first place? In the standard systems, there is no inference rule for introducing a free variable. Thus, they must come from axioms. And this is where the problem arises.

To see how, suppose that we are given a sequent calculus with a set of inferential rules including the ones mentioned above and a set of logical axioms. Also suppose that our language contains two predicates, say P and Q, standing for "is a bachelor" and "is unmarried," respectively. Given this translation, it must hold that for each individual constant a: $Pa \mathrel{\mid\!\sim} Qa$. So let us add these implications as proper atomic axioms (i.e., these are *material* inferences in which the meanings of "P" and "Q" partly consist). Now, given the standard right rule of the conditional, it follows that for each individual constant a: $\mathrel{\mid\!\sim} Pa \rightarrow Qa$. However, because in the standard setting any open sequents in which free eigenvariables occur must come from logical axioms (i.e., those implications that are formally valid irrespective of the expressions involved in them, such as tautologies [i.e., $A \mathrel{\mid\!\sim} A$] or the explosion [i.e., $\bot \mathrel{\mid\!\sim}$]), there is no way to derive the sequent in which the corresponding free eigenvariable occurs: $\mathrel{\mid\!\sim} Py \rightarrow Qy$. And because it is only via such an open sequent that we can derive the target sequent with the corresponding universal on the right—i.e., $\mathrel{\mid\!\sim} \forall x(Px \rightarrow Qx)$—, the left-to-right direction of the universal principle fails.

There are two possible routes to blocking this underproduction problem: adding more open sequents or modifying the right universal rule. One immediate response from the first route would simply be to add the required open sequent as an extra proper axiom: $Py \mathrel{\mid\!\sim} Qy$. After all, one may think, given that for any a, $Pa \mathrel{\mid\!\sim} Qa$ is materially good, $Py \mathrel{\mid\!\sim} Qy$ must also be materially good, and therefore count as a proper axiom. This option is, however, not available to us as logical inferentialists with respect to variables. Our aim is to *explain* the meaning of a variable in terms of the rules governing its inferential use. Yet, if we were simply to stipulate a crucial aspect of variable use as above in the form of a proper axiom, we would rather *exploit* our presupposed understanding of its meaning than *explain* that meaning. The aspect in question should be derived from some fundamental rule(s) governing the inferential use of the variable in question (i.e., "y") instead of being simply stipulated.

Before turning to my own solution, I should also mention the response from the second route. One may think that the underproduction problem can also be straightforwardly solved by letting the right rule for the universal "look directly at" the material inferences—the inferences in which the meanings of individual constants and predicates such as "a," "P," and "Q" (partly) consist. After all, if $\mathrel{\mid\!\sim} Pa \to Qa$ for an arbitrary a, it ipso facto seems plausible to derive $\mathrel{\mid\!\sim} \forall x(Px \to Qx)$. Thus, the following modified right universal rule may suggest itself:

$$\frac{\Gamma \mathrel{\mid\!\sim} A[\alpha/\xi], \Theta}{\Gamma \mathrel{\mid\!\sim} \forall \xi A, \Theta} \text{R}\forall'$$

where no α occurs in Γ or Θ,

where the modified eigenvariable condition at the bottom is supposed to ensure the "arbitrariness" of the relevant constant α. A serious problem with this solution, however, is that in the presence of proper axioms, the modified condition no longer ensures the arbitrary substitutability of α at all. For that matter, the eigenvariable condition and any of their variants can play its intended role only under the assumption that each term has exactly the same inferential potential (i.e., $\Gamma[\tau_1/\xi] \mathrel{\mid\!\sim} \Theta[\tau_1/\xi] \Leftrightarrow \Gamma[\tau_2/\xi] \mathrel{\mid\!\sim} \Theta[\tau_2/\xi]$). This holds as long as we limit our focus to logical inferences in which the inferential potential that is characteristic of individual constants is generally ignored; but that assumption may no longer hold once we take such potential into account by adding the relevant proper axioms. Suppose, for instance, that b, c, and R stand for "Tokyo," "Japan," and "is hit by a typhoon," respectively. Given this translation, the following implication seems to count as a proper axiom: $Rb \mathrel{\mid\!\sim} Rc$. However, in the presence of this axiom, the modified right universal rule presented above lets us derive an obviously invalid implication: $Rb \mathrel{\mid\!\sim} \forall x Rx$ (note that the modified eigenvariable condition is satisfied here as no c occurs in Rb).

In my view, the real culprit of the problem of underproduction is not the right universal rule, but the fact that the standard systems are devoid of any inferential rules that govern the use of variables in such a way as to make them mean what they mean—any inferential rules that let us derive the set of new open sequents that are not formally valid but materially good given the inferential potential of the other expressions (encapsulated in the set of proper axioms). In the next two sections, I demonstrate how such rules can be built into a supraintuitionistic sequent calculus that meets *Nonmonotonicity* and several conditions of *Expressivity*.

A First-Order Sequent Calculus for Logical Inferentialists/Expressivists

3 The system

Let $\mathcal{L}_0$ be the atomic language with the bottom ("$\perp$"), Var the set of all the variables, and Con the set of all the individual constants, where I assume that Con is finite. Let $\mathcal{L}_{0-} = \mathcal{L}_0 - \{\perp\}$, and $\mathcal{L}_{0[-]}^c$ is the largest closed subset of $\mathcal{L}_{0[-]}$ (i.e., $\mathcal{L}_{0[-]}^c = \{p \mid p \in \mathcal{L}_{0[-]}$, and p is a closed sentence$\}$. Note that the square brackets are used here to indicate that the bracketed element (i.e., "$-$") is optional (thus, here I set $\mathcal{L}_0^c$ and $\mathcal{L}_{0-}^c$ at a stroke). For notational convenience, I shall often use such square brackets.

Next, let $\mathrel{\mid\!\sim}_0$ be a material consequence relation over $\mathcal{L}_0^c$. Because $\mathrel{\mid\!\sim}_0$ is *material*, $\mathrel{\mid\!\sim}_0$ varies depending on what particular vocabulary $\mathcal{L}_0^c$ contains and how such vocabulary should be used in our discursive practice. Although it is an important task to think about how to identify $\mathrel{\mid\!\sim}_0$ within a given discursive practice, that is beyond and orthogonal to my purposes here. Thus, I simply take $\mathrel{\mid\!\sim}_0$ as given. I assume, though, that $\mathrel{\mid\!\sim}_0$ meets at least a few structural constraints, as follows:

Definition 1 $\mathrel{\mid\!\sim}_0 \subseteq \mathcal{P}(\mathcal{L}_{0-}^c) \times \mathcal{L}_0^c$, where (i) $\mathcal{L}_{0-}^c \mathrel{\mid\!\sim}_0 \perp$; (ii) $\emptyset \mathrel{\mid\!\not\sim}_0 \perp$; (iii) $\mathrel{\mid\!\sim}_0$ *is reflexive;* (iv) *if for any* $\Delta_0 \subseteq \mathcal{L}_{0-}^c : \Delta_0, \Gamma_0 \mathrel{\mid\!\sim}_0 \perp$, *then for any* $p \in \mathcal{L}_0^c : \Gamma_0 \mathrel{\mid\!\sim}_0 p$.

Note that (iv) is a weak version of the explosion principle, which I call, after Hlobil (2016), "Ex Falso *Fixo* Quodlibet" (ExFF). Because I treat the premises as a set, contraction and permutation are also granted. However, I do not impose weakening on this base consequence relation in order to make room for it to be defeasible (and therefore nonmonotonic). Transitivity is not imposed either, for the same reason.[9]

Now, let us turn to the extension of a material consequence relation thus defined over $\mathcal{L}_0^c$ (i.e., $\mathrel{\mid\!\sim}_0$) to the (indexed) consequence relation over the logically complex language $\mathcal{L}$ (i.e., $\mathrel{\mid\!\sim}_{[S]}^{[\uparrow X]}$). $\mathcal{L} = \mathcal{L}_- \cup \{\perp\}$, where the syntax of $\mathcal{L}_-$ is given as follows:

Syntax of $\mathcal{L}_-$: $\varphi ::= p \mid \varphi \to \varphi \mid \neg\varphi \mid \varphi \& \varphi \mid \varphi \vee \varphi \mid \Box\varphi \mid \forall x_i \varphi \mid \exists x_i \varphi$.

For technical convenience, I assume that variables in $\mathcal{L}$ consist of "x" with varying indices (i.e., $Var = \{x_1, x_2, \ldots x_i, \ldots\}$).

The extension of $\mathrel{\mid\!\sim}_0$ over $\mathcal{L}_0^c$ in this logically complex language is conducted via a sequent calculus, which I call the first-order nonmonotonic

---
[9] Transitivity forces monotonicity in the presence of the conditional satisfying the deduction theorem and its converse (Hlobil, 2016, sec. 4.3).

sequent calculus (FNM). Note that FNM not only specifies the snake turnstile ($\mathrel{\mid\kern-0.4em\sim}$) that represents the extended consequence relation, but also introduces those with indices ($\mathrel{\mid\kern-0.4em\sim}_S^{\uparrow X}$) in order to keep track of important features of $\mathrel{\mid\kern-0.4em\sim}$ concerning subjunctive robustness and universalizability of inferential pattern. The indexed upward-arrow is supposed to track a range of subjunctive robustness (i.e., $\Gamma \mathrel{\mid\kern-0.4em\sim}^{\uparrow X} A \Leftrightarrow$ for any $\Delta \in X \subseteq \mathcal{P}(\mathcal{L}_{0-}^c)$: $\Gamma, \Delta \mathrel{\mid\kern-0.4em\sim} A$), whereas the substitution memory S is supposed to register a range of variable substitutability *salva consequentia* (i.e., $\Gamma \mathrel{\mid\kern-0.4em\sim}_S A \Leftrightarrow$ for any $\alpha \in N \subseteq Con$ s.t. $< N, i > \in S$: $\Gamma[\alpha/x_i] \mathrel{\mid\kern-0.4em\sim} A[\alpha/x_i]$). Although FNM thus deals with countably many turnstiles, the main turnstile is the plain snake, within which several important features of the extended consequence relation (including those tracked by the two indices above) can eventually be expressed with the help of logical operators such as $\rightarrow$, $\neg$, $\Box$, and $\forall$ (see section 4.3).

First, a given material consequence relation is taken as the axioms of FNM.

Axiom: If $\Gamma \mathrel{\mid\kern-0.4em\sim}_0 p$, then $\Gamma \mathrel{\mid\kern-0.4em\sim}_\emptyset^{\uparrow\emptyset} p$ is an axiom.
Convention 1: $\mathrel{\mid\kern-0.4em\sim}_S^{\uparrow\emptyset}$ can be abbreviated as $\mathrel{\mid\kern-0.4em\sim}_S$.
Convention 2: $\mathrel{\mid\kern-0.4em\sim}_\emptyset^{[\uparrow X]}$ can be abbreviated as $\mathrel{\mid\kern-0.4em\sim}^{[\uparrow X]}$.
Convention 3: $\mathrel{\mid\kern-0.4em\sim}_{[S]}^{\uparrow \mathcal{P}(\mathcal{L}_{0-}^c)}$ can be abbreviated as $\mathrel{\mid\kern-0.4em\sim}_{[S]}^{\uparrow}$.

Then, FNM extends this base consequence relation by closing it under the following sequent rules.

$$\frac{\Gamma \mathrel{\mid\kern-0.4em\sim}_S^{\uparrow} A \quad \Gamma, B \mathrel{\mid\kern-0.4em\sim}_S^{\uparrow X} C}{\Gamma, A \rightarrow B \mathrel{\mid\kern-0.4em\sim}_S^{\uparrow X} C} \text{ LC} \qquad \frac{\Gamma, A \mathrel{\mid\kern-0.4em\sim}_S^{\uparrow X} B}{\Gamma \mathrel{\mid\kern-0.4em\sim}_S^{\uparrow X} A \rightarrow B} \text{ RC}$$

$$\frac{\Gamma \mathrel{\mid\kern-0.4em\sim}_S^{\uparrow X} A}{\Gamma, \neg A \mathrel{\mid\kern-0.4em\sim}_S^{\uparrow X} \bot} \text{ LN} \qquad \frac{\Gamma, A \mathrel{\mid\kern-0.4em\sim}_S^{\uparrow X} \bot}{\Gamma \mathrel{\mid\kern-0.4em\sim}_S^{\uparrow X} \neg A} \text{ RN}$$

$$\frac{\Gamma, A, B \mathrel{\mid\kern-0.4em\sim}_S^{\uparrow X} C}{\Gamma, A \& B \mathrel{\mid\kern-0.4em\sim}_S^{\uparrow X} C} \text{ L\&} \qquad \frac{\Gamma \mathrel{\mid\kern-0.4em\sim}_S^{\uparrow X} A \quad \Gamma \mathrel{\mid\kern-0.4em\sim}_S^{\uparrow X} B}{\Gamma \mathrel{\mid\kern-0.4em\sim}_S^{\uparrow X} A \& B} \text{ R\&}$$

$$\frac{\Gamma, A \mathrel{\mid\kern-0.4em\sim}_S^{\uparrow X} C \quad \Gamma, B \mathrel{\mid\kern-0.4em\sim}_S^{\uparrow X} C}{\Gamma, A \vee B \mathrel{\mid\kern-0.4em\sim}_S^{\uparrow X} C} \text{ LV} \qquad \frac{\Gamma \mathrel{\mid\kern-0.4em\sim}_S^{\uparrow X} A_i}{\Gamma \mathrel{\mid\kern-0.4em\sim}_S^{\uparrow X} A_1 \vee A_2} \text{ RV}$$
where $i = 1$ or 2

A First-Order Sequent Calculus for Logical Inferentialists/Expressivists

$$\frac{\Gamma, p_1, \ldots, p_n \mid\!\sim_S A \quad \Gamma \mid\!\sim_S^{\uparrow X} A}{\Gamma \mid\!\sim_S^{\uparrow\{\{p_1,\ldots,p_n\}\}\cup X} A} \text{PushUpUN}$$
where $p_1, \ldots, p_n \in \mathcal{L}_{0-}^c$

$$\frac{\Gamma, A \mid\!\sim_S^{\uparrow X} B}{\Gamma, \Box A \mid\!\sim_S^{\uparrow X} B} \text{LB} \qquad \frac{\Gamma \mid\!\sim_S^{\uparrow} A}{\Gamma \mid\!\sim_S^{[\uparrow]} \Box A} \text{RB}$$

$$\frac{\Gamma[\alpha/x_i] \mid\!\sim_S^{\uparrow X} A[\alpha/x_i]}{\Gamma \mid\!\sim_{S\cup\{<\{\alpha\},i>\}}^{\uparrow X} A} \text{AB1} \qquad \frac{\Gamma[\alpha/x_i] \mid\!\sim_S^{\uparrow X} A[\alpha/x_i]}{\Gamma \mid\!\sim_S^{\uparrow X} A} \text{AB2}$$
where there is no $<N, i> \in S$ $\qquad$ where there is no $<N, i> \in S$

$$\frac{\Gamma \mid\!\sim_{S\cup\{<N,i>\}}^{\uparrow X} A \quad \Gamma \mid\!\sim_{S\cup\{<N',i>\}}^{\uparrow X} A}{\Gamma \mid\!\sim_{S\cup\{<N\cup N',i>\}}^{\uparrow X} A} \text{vUN}$$

$$\frac{\Gamma, A \mid\!\sim_{S\cup\{<N,i>\}}^{\uparrow} B}{\Gamma, \forall x_i A \mid\!\sim_{S-\{<N,i>\}}^{[\uparrow]} B} \text{pLA} \qquad \frac{\Gamma \mid\!\sim_{S\cup\{<Con,i>\}}^{\uparrow X} A}{\Gamma \mid\!\sim_{S-\{<Con,i>\}}^{\uparrow X} \forall x_i A} \text{RA}$$

$$\frac{\Gamma, A \mid\!\sim_{S\cup\{<Con,i>\}}^{\uparrow X} B}{\Gamma, \exists x_i A \mid\!\sim_{S-\{<Con,i>\}}^{\uparrow X} B} \text{LE} \qquad \frac{\Gamma \mid\!\sim_{S\cup\{<N,i>\}}^{\uparrow X} A}{\Gamma \mid\!\sim_{S-\{<N,i>\}}^{\uparrow X} \exists x_i A} \text{RE}$$

$$\frac{\Gamma \mid\!\sim_S^{\uparrow} A}{\Gamma, B \mid\!\sim_S^{[\uparrow]} A} \text{pW} \qquad \frac{\Gamma \mid\!\sim_S^{\uparrow} \bot}{\Gamma \mid\!\sim_S^{[\uparrow]} A} \text{ExFF}$$

Most of these rules are adopted from the Non-Monotonic Modal sequent calculus in Hlobil (2016) (with obvious adjustments and a few minor modifications), for which I omit justifications. Among others, AB1 (for "abstraction") and vUN (for "variable unification") are crucial for the substitution memory to do its intended job, and given them RA is the key for the universal principle to hold. The top sequent of pLA must be indefeasible, since otherwise replacement of A by $\forall x_i A$ on the left might defeat the implication or incoherence at issue. In this way, FNM maps $\mid\!\sim_0 \subseteq \mathcal{P}(\mathcal{L}_{0-}^c) \times \mathcal{L}_0^c$ to $\mid\!\sim_S^{\uparrow X} \subseteq \mathcal{P}(\mathcal{L}_-) \times \mathcal{L}$, where $X \subseteq \mathcal{P}(\mathcal{L}_{0-}^c)$ and $S \subset \mathcal{P}(Con) \times \mathbb{N}$.

4 Main properties

4.1 Conservativeness

To begin with, it is straightforward to show that FNM is a conservative extension of the underlying material consequence relation, as there is no simplifying rule (such as *Cut*) in FNM.

Proposition 1 *For any $\Gamma_0 \subseteq \mathcal{L}_{0-}^c$ and for any $p \in \mathcal{L}_0^c$, $\Gamma_0 \mathrel{\mid\!\sim} p \Leftrightarrow \Gamma_0 \mathrel{\mid\!\sim_0} p$.*

Proof. ($\Rightarrow$) is straightforward. ($\Leftarrow$) is also straightforward from the fact that no rule of FNM reduces the complexity of a formula on either side of $\mid\!\sim$. □

As we stipulated that $\mid\!\sim_0$ can be nonmonotonic, so can $\mid\!\sim$, which is a conservative extension of $\mid\!\sim_0$. Thus, we have shown that FNM satisfies *Nonmonotonicity*.

4.2 Preservation of reflexivity

Next, reflexivity as stipulated at the base is preserved by FNM. With some preparations, the preservation of reflexivity is first shown with respect to the atomic *open* sentences.

Definition 2 *S is unique $\Leftrightarrow$ for any N such that $< N, i > \in S$, (i) $N \neq \emptyset$, and (ii) there is no N' such that $N' \neq N$ and $< N', i > \in S$.*

Lemma 1 *For any $\Gamma \subseteq \mathcal{L}_-$, for any $p \in \mathcal{L}_{0-}$, for any $X \subseteq \mathcal{P}(\mathcal{L}_{0-}^c)$, and for any finite unique $S \subset \mathcal{P}(Con) \times \mathbb{N}$, $\Gamma, p \mathrel{\mid\!\sim}_S^{\uparrow X} p$.*

Proof. By double induction on the number of distinct variables occurring in p and the cardinality of S. □

We are now in a position to show reflexivity across the board.

Proposition 2 *For any $\Gamma \subseteq \mathcal{L}_-$, for any $A \in \mathcal{L}_-$, and for any finite unique $S \subset \mathcal{P}(Con) \times \mathbb{N}$, $\Gamma, A \mathrel{\mid\!\sim}_S^{[\uparrow]} A$.*

Proof. By induction on the complexity of A. The base case is straightforward from Lemma 1, and the only tricky cases in the induction step are those in which A is the universal or the existential. Let A be $\forall x_i B$. By induction hypothesis, for any $a \in Con$, $B[a/x_i] \mathrel{\mid\!\sim}_S^{\uparrow} B[a/x_i]$, from which leaves $\forall x_i B \mathrel{\mid\!\sim}_S^{[\uparrow]} \forall x_i B$ is derivable via AB1, pLA, AB1, vUN ($|Con|-1$ times),

A First-Order Sequent Calculus for Logical Inferentialists/Expressivists

and RA. Note, however, that if for some $N, <N,i> \in S$, then the leaves must instead be $B[x_j/x_i][a/x_j] \mathrel{\vcenter{\hbox{$\sim$}}}_S^{[\uparrow]} B[x_j/x_i][a/x_j]$, where such j is chosen that x_j does not freely occur in B, and there is no $<N,j> \in S$ (otherwise, AB1 could not be applied). These leaves lead to $\forall x_j(B[x_j/x_i]) \mathrel{\vcenter{\hbox{$\sim$}}}_S^{[\uparrow]} \forall x_k(B[x_j/x_i])$, where $\forall x_j(B[x_j/x_i])$ is, by definition, syntactically equivalent to (or an "alphabetical variant" of) $\forall x_i B$. Finally, if needed, Γ can be added to the left by pW ($|\Gamma|$ times). The case in which A is the existential can be treated in the parallel manner. □

4.3 Expressivity

4.3.1 Implication and incoherence

According to logical expressivism, logical operators codify, within the object language, some features of the underlying material consequence relation. For instance, the conditional lets us talk about implications, whereas the negation lets us talk about incoherences. To justify such expressivist readings of logical operators in FNM, we first need to show the following lemma, which is repeatedly used in the proofs below.

Lemma 2 *If* $\Gamma \mathrel{\vcenter{\hbox{$\sim$}}}_S^{\uparrow X} A$, *then S is determinate.*

Proof. By induction on proof height. □

Now, the following proposition justifies a reading of the conditional and negation on the right as codifying implications and incoherences, respectively.

Proposition 3 *Conditional:* $\Gamma \mathrel{\vcenter{\hbox{$\sim$}}}_S^{\uparrow X} A \to B \Leftrightarrow \Gamma, A \mathrel{\vcenter{\hbox{$\sim$}}}_S^{\uparrow X} B$. **Negation:** $\Gamma \mathrel{\vcenter{\hbox{$\sim$}}}_S^{\uparrow X} \neg A \Leftrightarrow \Gamma, A \mathrel{\vcenter{\hbox{$\sim$}}}_S^{\uparrow X} \bot$. **Conjunction:** $\Gamma \mathrel{\vcenter{\hbox{$\sim$}}}_S^{\uparrow X} A \& B \Leftrightarrow \Gamma \mathrel{\vcenter{\hbox{$\sim$}}}_S^{\uparrow X} A$ *and* $\Gamma \mathrel{\vcenter{\hbox{$\sim$}}}_S^{\uparrow X} B$.

Proof. ($\Rightarrow$) is by induction on proof height. ($\Leftarrow$) is straightforward from RC, RN, and R&. □

4.3.2 Local monotonicity

Another important feature of the underlying material consequence relation that logical expressivists wish to codify is its local monotonicity. Although material inferences and incoherences often seem defeasible (e.g., "a is a match," "a is struck" $\vcenter{\hbox{$\sim$}}$ "a lights"), some may be *in*defeasible (e.g., "a is a bachelor" $\vcenter{\hbox{$\sim$}}$ "a is unmarried"). In FNM, such local monotonicity is kept track of by the upward-arrowed snake turnstile ($\mathrel{\vert\mkern-3mu\sim}^{\uparrow}$) and eventually expressed by the monotonicity box ($\Box$).[10] This claim is underwritten by the following lemmas and proposition.

Lemma 3 $\Gamma \mathrel{\vert\mkern-3mu\sim}_S^{\uparrow X} A \Leftrightarrow$ *for any* $\Delta \in X, \Gamma, \Delta \mathrel{\vert\mkern-3mu\sim}_S A$.

Proof. By induction on proof height. □

Lemma 4 $\Gamma \mathrel{\vert\mkern-3mu\sim}_S^{\uparrow X} \Box A \Leftrightarrow \Gamma \mathrel{\vert\mkern-3mu\sim}_S^{\uparrow} A$.

Proof. By induction on proof height. □

Proposition 4 $\Gamma \mathrel{\vert\mkern-3mu\sim}_S \Box A \Leftrightarrow$ *for any* $\Delta \subseteq \mathcal{L}_-, \Gamma, \Delta \mathrel{\vert\mkern-3mu\sim}_S A$.

Proof. Immediate from Lemma 3 and Lemma 4. □

4.3.3 Universalizability

Finally, let us turn to our key proposition, that is, the universal principle. We first need to do some preparation.

Sublemma 1 *If* $\tau_1 \neq \zeta, \tau_2 \neq \xi$, *and* $\xi \neq \zeta, A[\tau_1/\xi][\tau_2/\zeta] = A[\tau_2/\zeta][\tau_1/\xi]$.

Proof. By induction on the complexity of A. □

This lets us prove the following key lemma, which ensures that the substitution memory can store information about the range of variable substitutability *salva consequentia*.

Lemma 5 (i) *If* $\Gamma \mathrel{\vert\mkern-3mu\sim}_{S \cup \{<N,i>\}}^{\uparrow X} A$ *is proved at height* n, *then for any* $a \in N, \Gamma[a/x_i] \mathrel{\vert\mkern-3mu\sim}_{S-\{<N,i>\}}^{\uparrow X} A[a/x_i]$ *is provable at a height* $\leq n$. (ii) *If for any* $a \in N, \Gamma[a/x_i] \mathrel{\vert\mkern-3mu\sim}_{S-\{<N,i>\}}^{\uparrow X} A[a/x_i]$, *then* $\Gamma \mathrel{\vert\mkern-3mu\sim}_{S \cup \{<N,i>\}}^{\uparrow X} A$.

Proof. (i) is by induction on proof height, where Sublemma 1 is used for the cases of the induction step in which the sequent at issue comes by AB1 or AB2. (ii) is straightforward from AB1 and vUN. □

The next lemma says that the information about universal substitutability (i.e., a special case of the above) stored in the memory is expressible by the universal on the right.

---

[10]This technical apparatus was originally devised by Hlobil (2016).

A First-Order Sequent Calculus for Logical Inferentialists/Expressivists

Lemma 6 $\Gamma \mathrel{\vdash}^{\uparrow X}_{S-\{<Con,i>\}} \forall x_i A \Leftrightarrow \Gamma \mathrel{\vdash}^{\uparrow X}_{S \cup \{<Con,i>\}} A$, where x_i does not freely occur in Γ, and there is no $< N, i > \in S$.

Proof. ($\Rightarrow$) is by induction on proof height, where Lemma 5 (i) is used for the cases of the induction step in which the sequent at issue comes by pLA or LE. ($\Leftarrow$) is straightforward from RA. □

We are now in a position to prove the universal principle.

Proposition 5 $\Gamma \mathrel{\vdash} \forall x_i A \Leftrightarrow$ for any $a \in Con$, $\Gamma \mathrel{\vdash} A[a/x_i]$, where x_i does not freely occur in Γ.

Proof. Straightforward from Lemmas 5 and 6. □

4.4 Logical strength

Before closing, let us see where FNM is located within the familiar terrain of the other standard logics. FNM is *supraintuitionistic*. To demonstrate this, I first specify a special region of $\mathrel{\vdash}^{\uparrow X}_S$. Then, I embed intuitionistic logic there. Finally, I show that the special region is a local region of the material consequence relation represented by $\mathrel{\vdash}$.

Let us start with a few definitions.

Definition 3 *S is adequate for* $\Gamma \mathrel{\vdash}^{\uparrow X}_S A \Leftrightarrow$ for any x_i freely occurring in Γ or A, there is some N such that $< N, i > \in S$. *S is minimally adequate for* $\Gamma \mathrel{\vdash}^{\uparrow X}_S A \Leftrightarrow$ S is adequate, and if $< N, i > \in S$, then x_i freely occurs in Γ or A.

Definition 4 *S is full* $\Leftrightarrow$ for any $< N, i > \in S$, $N = Con$.

We are now in a position to specify a special region of $\mathrel{\vdash}^{\uparrow X}_S$ in which intuitionistic logic is going to be embedded.

Definition 5 $\Gamma \mathrel{\vdash}_F A \Leftrightarrow \Gamma \mathrel{\vdash}^{\uparrow}_S A$, where S is unique, minimally adequate, and full.

Next, let us show how to embed intuitionistic logic in $\mathrel{\vdash}_F$. This is done by showing that all the rules of LJ—the sequent calculus for first-order intuitionistic logic proposed by Gentzen (1935/1964)—can be added without disturbing $\mathrel{\vdash}_F$—that is, they are *admissible* within $\mathrel{\vdash}_F$. First, we need to define the notion of admissibility within $\mathrel{\vdash}_F$.

Definition 6 A rule R of the form: $\dfrac{\Gamma_1 \mathrel{\vert\kern-0.3ex\sim} A_1 \quad \cdots \quad \Gamma_n \mathrel{\vert\kern-0.3ex\sim} A_n}{\Gamma \mathrel{\vert\kern-0.3ex\sim} A}$ is admissible within $\mathrel{\vert\kern-0.3ex\sim}_F$ ⇔ if $\Gamma_1 \mathrel{\vert\kern-0.3ex\sim}_F A_1, \ldots, \Gamma_n \mathrel{\vert\kern-0.3ex\sim}_F A_n$, then $\Gamma \mathrel{\vert\kern-0.3ex\sim}_F A$.

Now, let us prove the admissibility of the rules of LJ within $\mathrel{\vert\kern-0.3ex\sim}_F$. It is convenient to discuss *Cut* and the rest of the rules separately. The admissibility of the latter can first be shown with the help of the following two sublemmas.

Sublemma 2 *All the connective rules concerning the propositional part of FNM (i.e., LC, RC, LN, RN, L&, R&, L∨, and R∨) are admissible within $\mathrel{\vert\kern-0.3ex\sim}_F$.*

Proof. RC, LN, RN, and L& are straightforward, whereas LC, R&, L∨, and R∨ are handled with the help of AB1 and vUN. □

Sublemma 3 *If $\Gamma \mathrel{\vert\kern-0.3ex\sim}_F A$, then for any $\Delta \subseteq \mathcal{L}_-$, $\Gamma, \Delta \mathrel{\vert\kern-0.3ex\sim}_F A$.*

Proof. Straightforward from Lemma 3. □

Lemma 7 *All the rules of LJ except for Cut are admissible within $\mathrel{\vert\kern-0.3ex\sim}_F$.*

Proof. All the rules except for those involving the quantifiers are straightforward from Sublemmas 2 and 3. The right universal and left existential rules are also straightforward from RA and LE respectively. Finally, the left universal and right existential rules are taken care of by pLA and RE with the help of Lemma 5. □

As to the admissibility of *Cut*, let us first prove the so-called "substitution lemma" with respect to $\mathrel{\vert\kern-0.3ex\sim}_F$. This is shown via the following sublemma.

Sublemma 4 *(i) If ζ does not freely occur in A, $A[\zeta/\xi][\tau/\zeta] = A[\tau/\xi]$. (ii) $A[\alpha/\xi][\alpha/\zeta] = A[\zeta/\xi][\alpha/\zeta]$.*

Proof. By induction on the complexity of A, where (i) is used in the inductive proof of (ii). □

Lemma 8 *If $\Gamma \mathrel{\vert\kern-0.3ex\sim}_F A$, then $\Gamma[\tau/x_i] \mathrel{\vert\kern-0.3ex\sim}_F A[\tau/x_i]$.*

Proof. By proof by cases. The cases are divided depending first on whether x_i freely occurs in Γ or A, then on whether τ is a variable, and, if so, further on whether τ freely occurs in Γ or A. In some of these cases, Lemma 5 and Sublemma 4 are appealed to. □

At this stage, we can appropriate Gentzen's (1935/1964) well-known result of the eliminability of *Cut* in LJ: given the substitution lemma, any sequents derivable in LJ via *Cut* are derivable without *Cut*.

Lemma 9 *Cut is also admissible within $\mathrel{\vdash}_F$.*

Proof. Straightforward from Lemma 7, Lemma 8, and the eliminability of *Cut* in LJ. □

Given these, we are now in a position to prove that $\mathrel{\vdash}_F$ is supraintuitionistic.

Lemma 10 *$\mathrel{\vdash}_F$ is at least as strong as intuitionistic logic.*

Proof. Straightforward from Proposition 2, Lemma 7, and Lemma 9. □

Finally, let us show that $\mathrel{\vdash}_F \subseteq \mathrel{\vdash}$, which immediately follows from the lemma below.

Lemma 11 *(i) If $\Gamma \mathrel{\vdash}_S^{\uparrow X} A$, then $\Gamma \mathrel{\vdash}^{\uparrow X} A$. (ii) If $\Gamma \mathrel{\vdash}^{\uparrow X} A$, then $\Gamma \mathrel{\vdash} A$.*

Proof. By induction on proof height. □

Corollary 1 *If $\Gamma \mathrel{\vdash}_F A$, then $\Gamma \mathrel{\vdash} A$.*

Thus, we are eventually in a position to show supraintuitionisity of $\mathrel{\vdash}$.

Proposition 6 *$\mathrel{\vdash}$ is at least as strong as intuitionistic logic.*

Proof. Straightforward from Lemma 10 and Corollary 1. □

5 Conclusion

This paper was motivated by two philosophical aims. One is to explain the meanings of variables in terms of their inferential roles; the other is to implement a logical system that can codify (i.e., talk about), within the object language, the inferential roles (i.e., meanings) of predicates. I argued that it is essential for both of these aims to have a proof system in which what I call the universal principle holds even in the presence of arbitrary nonlogical axioms. To the best of my knowledge, however, no currently available system meets this demand. Thus, I propose an alternative system in which the use of variables is explicitly controlled by an extra set of inferential rules in such a way that the universal principle holds robustly. This system is nonmonotonic and conservative, preserves reflexivity, and satisfies several biconditionals essential for its logical operators to codify important features of the underlying consequence relation, such as implications, incoherences, local monotonicity, and the universalizability of inferential patterns. The system is also supraintuitionistic.

References

Brandom, R. (1994). *Making It Explicit: Reasoning, Representing, and Discursive Commitment*. MA: Harvard University Press.

Brandom, R. (2000). *Articulating Reasons: An Introduction to Inferentialism*. MA: Harvard University Press.

Brandom, R. (2008). *Between Saying and Doing: Towards an Analytic Pragmatism*. Oxford: Oxford University Press.

Brandom, R. (2009). *Reason in Philosophy: Animating Ideas*. Cambridge, MA: Harvard University Press.

Fine, K. (1985). *Reasoning with Arbitrary Objects*. Oxford: Basil Blackwell.

Fine, K. (2007). *Semantic Relationism*. MA: Blackwell Publishing.

Frege, G. (1979). *Posthumous Writings* (P. Long & R. White, Trans.). Oxford: Basil Blackwell.

Gentzen, G. (1935/1964). Investigation into logical deduction (M. E. Szabo, Trans.). *American Philosophical Quarterly*, *1*(4), 288–306.

Hlobil, U. (2016). A nonmonotonic sequent calculus for inferentialist expressivists. In P. Arazim & M. Dančák (Eds.), *The Logica Yearbook 2015* (pp. 87–105). College Publications.

Kaplan, D. (2018). A multi-succedent sequent calculus for logical expressivists. In P. Arazim & M. Dančák (Eds.), *The Logica Yearbook 2017*. College Publications.

Peregrin, J. (2014). *Inferentialism: Why Rules Mater*. New York: Palgrave Macmillan.

Russell, B. (1905). On denoting. *Mind*, *14*(56), 479–493.

Russell, B. (1994). *Foundations of Logic 1903-05* (A. Urquhart, Ed.). London: Routledge.

Sellars, W. (1953). Inference and meaning. *Mind*, *62*(247), 313–338.

Shuhei Shimamura
Nihon University, College of Science and Technology
Japan
E-mail: `shuhei.shimamura0803@gmail.com`

A Dynamic Epistemic Logic for Resource-Bounded Agents

ANTHIA SOLAKI[1]

Abstract: In this paper, we present a dynamic epistemic logic suitable for resource-bounded agents. Our setting is informed by empirical evidence on deductive reasoning performance and therefore it avoids the problem of logical omniscience. In particular, we introduce actions capturing how the agent learns, forgets, and applies inference rules. Our model is a variant of Kripke models extended with impossible worlds and our updates modify its components (epistemic accessibility, rule availability, cognitive capacity) according to each action's effect. We further provide a sound and complete axiomatization, through a method connecting this semantic approach to logical omniscience with more syntactically-oriented ones. Finally, we use similar tools to model moderate introspective ability and thus avoid the unrealistic commitment to unbounded introspection.

Keywords: dynamic epistemic logic, logical omniscience, bounded rationality, resource-bounded agents, impossible worlds, rule-based agents

1 Introduction

The work of Hintikka (1962) paved the way for the formal study of propositional attitudes via possible-worlds semantics. Still, S5 modal logic, seen as the standard epistemic logic, faces *the problem of logical omniscience* (Fagin, Halpern, Moses, & Vardi, 1995): agents are modelled as reasoners with unbounded inferential power, always knowing anything that logically follows from what they know. For example, all tautologies are known regardless of their complexity, knowledge is closed under logical equivalence etc. However, it takes time to compute whether a complex formula is indeed a tautology; empirical evidence (e.g., on *framing effects*, Tversky & Kahneman, 1981) shows that logically equivalent statements are assessed differently by subjects, depending on their presentation. These examples suggest that cognitive effort involved in reasoning tasks needs to be accounted for.

---

[1] This work is funded by the Dutch Science Foundation, under the "PhDs in the Humanities" scheme (project number 322-20-018).

The S5 approach is sometimes defended due to its normative status; it models how we *ought to* reason. Yet experimental evidence suggests that mistakes in deductive reasoning are in fact systematic (Stanovich & West, 2000; Stenning & van Lambalgen, 2008). For example, people's performance in the *Wason Selection Task*[2], which essentially requires an application of modus ponens and modus tollens, is notoriously poor (Wason, 1966). Similarly, cognitive limitations affect the extent of introspection one can achieve. Apart from philosophical objections (Stroll, 1967; Williamson, 2000), the S5 axioms of *positive* and *negative* introspection ($K\phi \to KK\phi$ and $\neg K\phi \to K\neg K\phi$, respectively) are not in agreement with experimental findings (Verbrugge, 2009). We therefore seek another normative model tailored to such observations. This is why we propose modelling how a rational agent comes to know things, informed by empirical facts to ensure that "ought" actually implies "can".

To that end, we emphasize that although we are fallible and non-omniscient humans, we still are *logically competent*. Despite our failures, we still engage in bounded reasoning and introspection. As a result, we ask that agents should know those consequences that can be feasibly reached from their current epistemic state. Descriptive facts, e.g., regarding limitations of time and memory, are instrumental in determining the extent of feasibility.

The problem of logical omniscience has generally attracted much attention. One of the early suggestions (Hintikka, 1975) was to supplement possible-worlds models with *impossible* (or *non-normal*) worlds, that are not logically closed. If these are accessible to the agent, and given the common interpretation of knowledge as quantifying over accessible worlds, the closure properties for knowledge are invalidated. Still, such approaches are criticized on grounds of logical competence and explanatory power. Towards this direction, there are attempts discerning explicitly and implicitly held attitudes. In (Fagin & Halpern, 1987), agents have to be additionally *aware* of something to know it explicitly; hence their models are augmented by an *awareness function*, yielding the formulas the agent is aware of. However, aspects of the problem can be retained, logical competence may be sidestepped and it is not clear how the crucial notion of resource-boundedness can be accounted for. In (Quesada, 2009), the awareness set can be modified depending on the agent's applications of inference rules.

---

[2] The subject is given four cards, which have a number on one side and a color on the other. The visible sides of the cards are 3, 8, red and brown. The question of the task is: which cards must you turn over to test whether the proposition "if a card shows an even number on one side, then its opposite side is red" is true?

A Dynamic Epistemic Logic for Resource-Bounded Agents

Bjerring and Skipper (2018) provide an impossible-worlds framework that also focuses on reasoning steps the agent takes to come to know more.

While we build on the above-mentioned attempts, we are interested in capturing how resource-boundedness affects the reasoning processes underpinning knowledge; in particular, we use Dynamic Epistemic Logic (DEL) (Baltag & Renne, 2016; van Benthem, 2011; van Ditmarsch, van der Hoek, & Kooi, 2007) to keep track of the reasoning steps available to the agent, their (orderly) applications, and the cognitive effort required. This attempt is first presented in Sect. 2, dealing with deductive reasoning alone, thereby proposing a way out of logical omniscience. In Sect. 3, we give a sound and complete axiomatization, via a method that further allows for comparative remarks between syntactic and semantic approaches to the problem. This is followed by an extension of this framework towards a balanced view to introspection, given in Sect. 4.[3]

2 Resource-bounded deductive reasoning

2.1 Syntax

To begin with, we need a logical language where the rules of deductive reasoning are explicitly introduced. This is why we define:

Definition 1 (Inference rule) *Given $\phi_1, \ldots, \phi_n, \psi \in \mathcal{L}_P$, where $\mathcal{L}_P$ is the propositional language based on a set of atoms Φ, an inference rule ρ is a formula of the form $\{\phi_1, \ldots, \phi_n\} \rightsquigarrow \psi$.*

We denote the set of premises and the conclusion of ρ with $pr(\rho)$ and $con(\rho)$, while $\mathcal{L}_R$ denotes the set of all inference rules. However, since we focus on an agent's *knowledge*, we are interested in *truth-preserving* rules.

Definition 2 (Translation) *The translation of a rule ρ is given by $tr(\rho) := \bigwedge_{\phi \in pr(\rho)} \phi \rightarrow con(\rho)$.*

Then the definition of our framework's logical language is given by:

Definition 3 (Language) *First, we fix a set of constants $T := \{c_\rho \mid \rho \in \mathcal{L}_R\} \cup \{cp\}$. Then the language $\mathcal{L}$ is built as follows:*

$$\phi ::= z_1 s_1 + \ldots + z_n s_n \geq z \mid p \mid \neg\phi \mid \phi \wedge \phi \mid K\phi \mid A\rho \mid [+\rho]\phi \mid [-\rho]\phi \mid \langle\rho\rangle\phi$$

where $z_1, \ldots, z_n \in \mathbb{Z}$, $z \in \mathbb{Z}^r$, $s_1, \ldots, s_n \in T$, $p \in \Phi$, and $\rho \in \mathcal{L}_R$

---
[3] This paper is part of a recently initiated line of work, using DEL to model reasoning processes and resource-bounded agents (Berto, Smets, & Solaki, 2018; Smets & Solaki, 2018).

So the language is an extension of that of standard epistemic logic with:

- Numerical inequalities introduced to deal with cognitive effort (e.g., of the form $s_1 \geq s_2$). As we will see, this is possible because the constants of T essentially express the cognitive costs of inference rules and the agent's cognitive capacity.[4]

- A, an operator introduced to capture the agent's availability of inference rules. Specifically, $A\rho$ is to say that ρ is available to the agent, who can therefore apply it.

- Dynamic operators of the form $[+\rho]$ (resp. $[-\rho]$), such that: $[+\rho]\phi$ (resp. $[-\rho]$) says "after the agent learns (resp. forgets) ρ, ϕ is true".

- Dynamic operators of the form $\langle \rho \rangle$, such that: $\langle \rho \rangle \phi$ stands for "after applying ρ, ϕ is true".

2.2 Semantics: defining models

Our semantic model makes use of *impossible worlds*. Here, we abide by the so-called *American stance* (Berto, 2013): impossible worlds are not closed under *any* notion of logical consequence. Still, we want to build a model respecting the *minimal rationality* of agents. According to Cherniak (1986) we need a "theory of feasible inferences" where the difficulty of deductive reasoning is responsible for the agent performing *some*, but not all appropriate inferences, so in fact, we need a "well-ordering of inferences" in terms of difficulty. It is natural to connect this with the consumption of cognitive resources and use it to determine where the cutoff of an inferential chain lies. To start with, we rule out what is an obvious case of inconsistency for any logically competent agent: explicit contradictions. As the cognitive load increases while deductive reasoning evolves, we need to keep track of the cognitive costs of rules, with respect to each resource, having determined beforehand the resources (e.g., memory, attention, time etc.) considered. This is so because not all inference rules require equal cognitive effort, as indicated by experimental evidence (Johnson-Laird, Byrne, & Schaeken, 1992; Rips, 1994; Stenning & van Lambalgen, 2008). The cognitive effort will be captured by a (partial) function $c : \mathcal{L}_R \to \mathbb{N}^r$, where r is the number of resources considered. This function assigns a cost to each (sound) inference rule w.r.t. each resource. In addition, we will introduce *cognitive capacity*

---

[4]The choice of r, appearing in the definition of inequalities, will be made clear shortly after.

A Dynamic Epistemic Logic for Resource-Bounded Agents

in our model, a component expressing what the agent can afford w.r.t. each resource, meant to be decreased following each rule application.

Definition 4 (Semantic model) *Given a set of r-many resources Res, a model is a tuple $M = \langle W^P, W^I, f, V, R, cp \rangle$ where:*

- *W^P, W^I are non-empty sets of possible and impossible worlds respectively. Let $W := W^P \cup W^I$.*

- *$f : W \to \mathcal{P}(W)$ is a function mapping each world to its set of epistemically accessible worlds.*

- *$V : W \to \mathcal{P}(\mathcal{L})$ is a function mapping each world to a set of formulas. In possible worlds, the function intuitively assigns the set of atomic formulas true at the world. In impossible worlds, the function assigns all formulas true at the world.[5]*

- *$R : W \to \mathcal{P}(\mathcal{L}_R)$ is a function yielding the rules the agent has available (i.e., has acknowledged as truth-preserving) at each world.*

- *cp denotes the agent's cognitive capacity, i.e., $cp \in \mathbb{Z}^r$, intuitively standing for what the agent can afford w.r.t. each resource.*

In accordance to the remarks made above, we ask that: [6]
Minimal Consistency (MC): $\{\phi, \neg\phi\} \not\subseteq V(w)$ for all $w \in W^I, \phi \in \mathcal{L}$
Soundness of rules (SoR): for $w \in W^P$, if $\rho \in R(w)$ then $M, w \models tr(\rho)$

2.3 Logical dynamics: learning, forgetting and applying a rule

The language contains operators for actions capable of changing the rules that are available to the agent as well as her epistemic state following a certain rule-application. The semantic effect of these actions is, as usual in DEL, captured via *model transformations*. If a formula is of the form $[]\phi$ with $[]$ such an operator, then it is evaluated in a model by examining what the truth value of ϕ is at the transformed model.

The model transformation due to learning (resp. forgetting) ρ is obtained by suitably expanding (resp. restricting) the relevant model component.

---

[5] For simplicity, we view the valuation function as $V := V_p \cup V_i$, where the functions V_p and V_i that take care of possible and impossible worlds are injective.

[6] Assuming that propositional formulas are evaluated as usual at possible worlds.

Definition 5 (Model transformation by learning a rule) *Given a model M, its transformation by learning a rule ρ is a model $M^{+\rho}$ with*

$$R^{+\rho}(w) = \begin{cases} R(w) \cup \{\rho\} & \text{if } \rho \text{ is sound} \\ R(w) & \text{otherwise} \end{cases}$$

for all $w \in W^P$. Everything else remains as in the original model.

Definition 6 (Model transformation by forgetting a rule) *Given a model M, its transformation by forgetting a rule ρ is a model $M^{-\rho}$ with $R^{-\rho}(w) = R(w) \setminus \{\rho\}$, for all $w \in W^P$. Everything else remains as in the original.*

To capture the change induced by applications of inference rules, we have to encode them on the structure of our models. The effect of applying a rule is an expansion of the agent's information. We first introduce the notation $V^*(w)$ to restrict $V(w)$ to the propositional formulas satisfied at world w. We then impose the following condition to ensure that there are worlds capable of representing such expansions:

Succession: For every $w \in W$, if: (a) $pr(\rho) \subseteq V^*(w)$, (b) $\neg con(\rho) \notin V^*(w)$, and (c) $con(\rho) \neq \neg \phi$ for all $\phi \in V^*(w)$, then there is some $u \in W$ such that $V^*(u) = V^*(w) \cup \{con(\rho)\}$. We call u a ρ-expansion.

Next, we define the *ρ-radius*, in order to represent how ρ triggers an informational change, to the extent that *Minimal Consistency* is respected.

Definition 7 (ρ-radius) *We define the ρ-radius of a world w as follows:*

$$w^\rho := \begin{cases} \{w\}, \text{ if } pr(\rho) \nsubseteq V^*(w) \\ \emptyset, \text{ if } pr(\rho) \subseteq V^*(w) \text{ and} \\ \quad (\neg con(\rho) \in V^*(w) \text{ or } con(\rho) = \neg\phi \text{ for some } \phi \in V^*(w)) \\ \{u \mid u \text{ is a } \rho\text{-expansion of } w\}, \text{ if } pr(\rho) \subseteq V^*(w) \text{ and} \\ \quad \neg con(\rho) \notin V^*(w) \text{ and } con(\rho) \neq \neg\phi \text{ for all } \phi \in V^*(w) \end{cases}$$

Notice that w's ρ-radius amounts to $\{w\}$ for $w \in W^P$, due to the closure of possible worlds, while the radius of an impossible world may contain another impossible world. The radius is instrumental in modifying epistemic accessibility in the transformed model, after a rule-application.

Definition 8 (Model transformation by application of a rule) *Take $M = \langle W^P, W^I, f, V, R, cp \rangle$ and $w \in W^P$. The transformation of the pointed model (M, w) by an application of ρ is the pointed model (M^ρ, w) with*

A Dynamic Epistemic Logic for Resource-Bounded Agents

- $W^\rho = W$, $V^\rho = V$, $R^\rho = R$, and $cp^\rho = cp - c(\rho)$.

- $f^\rho = g$ such that $g(v) = \begin{cases} \bigcup_{u \in f(w)} u^\rho, & \text{for } v = w \\ f(v), & \text{for } v \neq w \end{cases}$

That is, (M^ρ, w) is obtained by (a) replacing w's epistemically accessible worlds in M with the elements of their ρ-radii, and (b) reducing the cognitive capacity by the cost of performing the ρ-step. Notice that the properties of our models are preserved under the three operations defined.

2.4 Truth clauses

In order to give the truth clauses for our formulas, we first need to assign interpretations to the constants in T:

Definition 9 (Interpretation of terms) *Given* $M = \langle W^P, W^I, f, V, R, cp \rangle$ *parameterized by resources Res and the cognitive cost function c, the constants of T are interpreted as follows:* $cp^M = cp$ *and* $c_\rho^M = c(\rho)$.

Definition 10 (Truth clauses) *The clauses below define when a formula is true at w in M. For* $w \in W^I$: $M, w \models \phi$ *iff* $\phi \in V(w)$. *For* $w \in W^P$:

$M, w \models z_1 s_1 + \ldots + z_n s_n \geq z$ iff $z_1 s_1^M + \ldots + z_n s_n^M \geq z$
$M, w \models p$ iff $p \in V(w)$
$M, w \models \neg\phi$ iff $M, w \not\models \phi$
$M, w \models \phi \land \psi$ iff $M, w \models \phi$ and $M, w \models \psi$
$M, w \models A\rho$ iff $\rho \in R(w)$
$M, w \models [+\rho]\phi$ iff $M^{+\rho}, w \models \phi$
$M, w \models [-\rho]\phi$ iff $M^{-\rho}, w \models \phi$
$M, w \models K\phi$ iff $M, w' \models \phi$ for all $w' \in f(w)$
$M, w \models \langle\rho\rangle\phi$ iff $M, w \models cp \geq c_\rho$, $M, w \models A\rho$ and $M^\rho, w \models \phi$

Notice that our intended reading of $\geq$ is that, for example, $s_1 \geq s_2$ iff every i-th component of s_1 is greater or equal than the i-th component of s_2. Validity is defined with respect to possible worlds only. Given our clause for K, the presence of impossible worlds, where formulas are assigned a truth value directly rather than recursively, suffices to break the closure principles of logical omniscience. On the other hand, despite being fallible, an agent can still come to know consequences of her knowledge and gradually eliminate impossibilities she initially entertained. Consider the truth conditions for epistemic assertions like $K\phi$ prefixed by a rule ρ; they require that

(a) the rule is executable, (b) the rule is available to the agent, (c) ϕ follows from the accessible worlds via an application of ρ. Moreover, the actions of learning and forgetting rules can affect the availability, and thus account for the flexibility of actual reasoning and its possible failures. Cognitive capacity, decreasing suitably after every rule-application, determines to which extent consequences of one's knowledge can come to be in turn known. This cutoff is therefore cognitively informed and not arbitrarily fixed. Overall, this approach encompasses each rule's different contribution and effort, and explains how resources are consumed as reasoning evolves.

Example 1 Consider the following scenario: agent Alice is given 2 cards, each has a number on one side and a color (red or brown) on the other. Alice knows that if a card has an *even* number on one side, it has *red* on the other. Suppose that the 1st card has 8 on its visible side (fact denoted by e_1), and the 2nd card has brown (denoted by $\neg r_2$; r_2 stands for "the second card is *red*"). What should the agent derive? In seeing the even card and performing a modus ponens (MP) step (if affordable), she comes to know r_1. In seeing the brown card and performing a modus tollens (MT) step (if affordable), she comes to know $\neg e_2$. Figure 1 shows the original and the MP, MT-updated model. However, empirical evidence suggests that MT is more cognitively costly than MP (Johnson-Laird et al., 1992; Rips, 1994; Stenning & van Lambalgen, 2008). So the cost of MT exceeds the cost of MP, and it might be the case that it is so cognitively costly for Alice to apply that she cannot do so, e.g., under time pressure (consider subjects given the Wason Selection Task to complete in specific time bounds). Moreover, as many argue that MT, unlike MP, is not a primitive rule (Rips, 1994), it might be that the rule is not even available to Alice, so she needs to *learn* it and then apply it. Such scenarios exemplify how our tools can fit reasoning tasks studied in psychology of reasoning.

3 Axiomatization

3.1 Semantic and syntactic approaches

We now put forward a reduction of impossible-worlds models to Kripke models (i.e., involving solely possible worlds) augmented by syntactic functions. These functions capture the effect of impossible worlds in epistemic accessibility. In this way, we wish to combine the fallibility of real agents, for which impossible worlds stand, and the simplicity in technical manipu-

A Dynamic Epistemic Logic for Resource-Bounded Agents

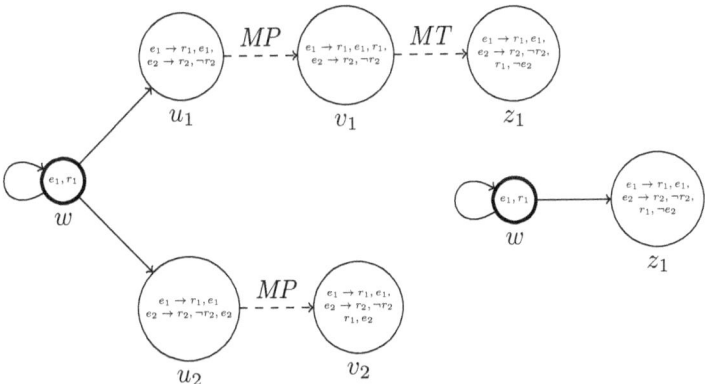

Figure 1: For worlds in W^P, unlike those in W^I, we use thicker nodes and write down only the atomic formulas satisfied there. Dashed lines indicate rule expansions. Left: the original model for Alice, who has not yet derived all consequences of her knowledge, entertaining an incomplete (u_1) and an inconsistent (u_2) world. Right: the updated model, following the applications of MP, MT.

lation of structures resembling those of Fagin and Halpern (1987). Besides, a rough division of attempts against logical omniscience is between syntactic and semantic ones. According to Fagin and Halpern (1987), a syntactic approach lacks the elegance of a semantic (impossible-worlds) one, but the latter's semantic rules do not adequately capture intuitions about knowledge and belief. To increase explanatory power and intuitiveness, we limited the arbitrariness of impossible worlds (via *Succession* and *Minimal Consistency*) and modelled logically competent agents, in that they gradually refine their epistemic state through rule-applications. In what follows, we reduce our models to syntactic structures to extract a sound and complete logic. In combination with Wansing (1990), it can be claimed that there is a "correspondence" between the two styles of attacking logical omniscience.

3.2 A common background language $\mathcal{L}_r$

In order to show that the same formulas are valid under the original and the reduced models, we introduce auxiliary operators to the *static* fragment of $\mathcal{L}$. These are such to discern the impact of possible and impossible worlds and to encode our model's structure. These operators, along with their interpretation at possible worlds, are given below:

$M, w \models L\phi$ iff $M, u \models \phi$ for all $u \in W^P \cap f(w)$
$M, w \models I\phi$ iff $M, u \models \phi$ for all $u \in W^I \cap f(w)$
$M, w \models \hat{I}\phi$ iff $M, u \models \phi$ for some $u \in W^I \cap f(w)$
$M, w \models \langle RAD \rangle_\rho \phi$ iff for some $u \in w^\rho$: $M, u \models \phi$[7]

Abbreviations: (a) $[RAD]_\rho \phi := \langle RAD \rangle_\rho \top \to \langle RAD \rangle_\rho \phi$, and (b) If ϕ is of the form $\neg \psi$, for some formula ψ, then $\overline{I}\phi := \hat{I}\psi$, else $\overline{I}\phi := \bot$.

3.3 The reduced model

Take a model $M = \langle W^P, W^I, f, V, R, cp \rangle$ and $V_r(w) := \{\phi \in \mathcal{L}_r \mid M, w \models \phi\}$, for $w \in W^I$. We construct $\mathfrak{M} = \langle W, f, V, I, \hat{I}, R, cp \rangle$ where:

- $W = W^P$, $f(w) = f(w) \cap W$ for $w \in W$, $V = V|_W$, and $R = R|_W$.

- $I: W \to \mathcal{P}(\mathcal{L}_r)$ such that $I(w) = \bigcap_{v \in f(w) \cap W^I} V_r(v)$. Intuitively, I takes a possible world w and yields the set of those formulas that are true at all impossible worlds accessible from w.

- $\hat{I}: W \to \mathcal{P}(\mathcal{L}_r)$ such that $\hat{I}(w) = \bigcup_{v \in f(w) \cap W^I} V_r(v)$. Intuitively, $\hat{I}$ takes a possible world w and yields the set containing any formula true at some impossible world accessible from w.

The interpretation of terms in $\mathfrak{M}$ is as before, for it depends on the parameter c and the model component cp. The semantics based on $\mathfrak{M}$ is:

$\mathfrak{M}, w \models z_1 s_1 + \ldots + z_n s_n \geq z$ iff $z_1 s_1^\mathfrak{M} + \ldots + z_n s_n^\mathfrak{M} \geq z$
$\mathfrak{M}, w \models p$ iff $p \in V(w)$
$\mathfrak{M}, w \models L\phi$ iff for all $u \in f(w)$: $\mathfrak{M}, u \models \phi$
$\mathfrak{M}, w \models \neg \phi$ iff $\mathfrak{M}, w \not\models \phi$
$\mathfrak{M}, w \models I\phi$ iff $\phi \in I(w)$
$\mathfrak{M}, w \models \phi \land \psi$ iff $\mathfrak{M}, w \models \phi$ and $\mathfrak{M}, w \models \psi$
$\mathfrak{M}, w \models \hat{I}\phi$ iff $\phi \in \hat{I}(w)$
$\mathfrak{M}, w \models A\phi$ iff $\phi \in R(w)$
$\mathfrak{M}, w \models K\phi$ iff $\mathfrak{M}, w \models L\phi$ and $\mathfrak{M}, w \models I\phi$
$\mathfrak{M}, w \models \langle RAD \rangle_\rho \phi$ iff for some $u \in w^\rho$: $\mathfrak{M}, u \models \phi$

Theorem 1 (Reduction) *Given a model M, construct $\mathfrak{M}$ as described above. Then $\mathfrak{M}$ is a reduction of M, i.e., for any $w \in W^P$ and formula $\phi \in \mathcal{L}_r$: $M, w \models \phi$ iff $\mathfrak{M}, w \models \phi$.*

A Dynamic Epistemic Logic for Resource-Bounded Agents

Proof. By induction on the complexity of ϕ. Recall that validity is defined with respect to possible worlds in the original model. The base case, and the steps for inequalities, $\neg$, $\wedge$ and A are straightforward. For $L, I, \hat{I}, \langle RAD \rangle_\rho$ we rely on the construction of the auxiliary operators and the definition of $\mathfrak{M}$. For K, the claim holds as it can be re-expressed in terms of L and I. □

3.4 Static axiomatization and reduction axioms

We first present an axiomatic system for the static part and show that it is sound and complete w.r.t. the reduced Kripke models. For the dynamic part, involving $\langle \rho \rangle, [+\rho], [-\rho]$, we give *reduction axioms*. As usual in DEL, the static logic combined with these axioms, suffices to get a complete logic for our full framework.

Definition 11 (Axiomatization) *The static logic is axiomatized by the following axioms and the rules Modus Ponens and Necessitation (from ϕ infer $L\phi$).*

PC	All instances of classical propositional tautologies
$Ineq$	All instances of valid formulas about linear inequalities
K	$L(\phi \to \psi) \to (L\phi \to L\psi)$
T	$L\phi \to \phi$
MC	$\neg(I\phi \wedge I\neg\phi)$
SoR	$A\rho \to tr(\rho)$
$Succession$	$(\bigwedge_{\psi \in pr(\rho)} I\psi \wedge \neg\hat{I}\neg con(\rho) \wedge \neg\overline{I} con(\rho)) \to I\langle RAD \rangle_\rho con(\rho) \wedge$
	$(I\phi \to I\langle RAD \rangle_\rho \phi)$, for $\phi \in \mathcal{L}_P$
	$I\langle RAD \rangle_\rho \phi \to I\phi$, for $\phi \in \mathcal{L}_P$ and $\phi \neq con(\rho)$
	$\neg \bigwedge_{\psi \in pr(\rho)} I\psi \to (I\phi \leftrightarrow I\langle RAD \rangle_\rho \phi)$, for $\phi \in \mathcal{L}_P$
	$\bigwedge_{\psi \in pr(\rho)} I\psi \wedge (\hat{I}\neg con(\rho) \vee \overline{I} con(\rho)) \to I[RAD]_\rho \bot$
Red_1	$K\phi \leftrightarrow (L\phi \wedge I\phi)$
Red_2	$\langle RAD \rangle_\rho \phi \leftrightarrow \phi$
Red_3	$I[RAD]_\rho \phi \leftrightarrow (I\langle RAD \rangle_\rho \top \to I\langle RAD \rangle_\rho \phi)$

Ineq, described by Fagin and Halpern (1994), is introduced to deal with inequalities.[8] *MC*, *SoR* and *Succession* correspond to our model conditions, given how these are reflected on our language. We use T too, because this corresponds to factivity of knowledge. Red_1 reduces K in terms of L and I. Red_2 and Red_3 capture the properties of ρ-expansions. Using **M** for the class of reflexive $\mathfrak{M}$ models we show:

---
[8]Of course, the axioms in *Ineq* are adapted because terms are interpreted as r-tuples.

Anthia Solaki

Theorem 2 (Soundness, Completeness) *The static logic is sound and complete w.r.t.* **M**.

Proof. Soundness: It suffices to show that the reduction axioms are valid in this class. *Completeness*: We need to construct a suitable canonical model, corresponding to our $\mathfrak{M}$ models and their properties. This can be defined as $\mathcal{M} = \langle \mathcal{W}, \mathcal{F}, \mathcal{V}, \mathcal{I}, \hat{\mathcal{I}}, \mathcal{R}, cp \rangle$, where $\mathcal{W}, \mathcal{F}, \mathcal{V}$ are defined as usual (Blackburn, de Rijke, & Venema, 2001). Its functions are given by: $\mathcal{I}(w) = \{\phi \mid I\phi \in w\}$, $\hat{\mathcal{I}}(w) = \{\phi \mid \hat{I}\phi \in w\}$ and $\mathcal{R}(w) = \{\rho \mid A\rho \in w\}$, with $w \in \mathcal{W}$. Then we show the truth lemma (i.e., $\mathcal{M}, w \models \phi$ iff $\phi \in w$) by induction on ϕ. Completeness follows by standard modal logic results. □

Before we move to reduction axioms for our three actions, we have to express the updated terms in the language: $cp^\rho := cp - c_\rho$ and $c_\rho^\rho := c_\rho$.

Theorem 3 (Reduction axioms) *The following are valid in* **M**:

$\langle \rho \rangle (z_1 s_1 + \ldots + z_n s_n \geq z) \leftrightarrow (cp \geq c_\rho) \wedge A\rho \wedge (z_1 s_1^\rho + \ldots + z_n s_n^\rho \geq z)$
$\langle \rho \rangle p \leftrightarrow (cp \geq c_\rho) \wedge A\rho \wedge p$
$\langle \rho \rangle \neg \phi \leftrightarrow (cp \geq c_\rho) \wedge A\rho \wedge \neg \langle \rho \rangle \phi$
$\langle \rho \rangle (\phi \wedge \psi) \leftrightarrow (cp \geq c_\rho) \wedge A\rho \wedge \langle \rho \rangle \phi \wedge \langle \rho \rangle \psi$
$\langle \rho \rangle L\phi \leftrightarrow (cp \geq c_\rho) \wedge A\rho \wedge L\phi$
$\langle \rho \rangle I\phi \leftrightarrow (cp \geq c_\rho) \wedge A\rho \wedge I[RAD]_\rho \phi$
$\langle \rho \rangle K\phi \leftrightarrow (cp \geq c_\rho) \wedge A\rho \wedge K[RAD]_\rho \phi$
$\langle \rho \rangle A\sigma \leftrightarrow (cp \geq c_\rho) \wedge A\rho \wedge A\sigma$
$\langle \rho \rangle \hat{I}\phi \leftrightarrow (cp \geq c_\rho) \wedge A\rho \wedge \hat{I}\langle RAD \rangle_\rho \phi$
$\langle \rho \rangle \langle RAD \rangle_\rho \phi \leftrightarrow (cp \geq c_\rho) \wedge A\rho \wedge \langle RAD \rangle_\rho \phi$

$[+\rho](z_1 s_1 + \ldots + z_n s_n \geq z) \leftrightarrow (z_1 s_1 + \ldots + z_n s_n \geq z)$
$[+\rho]p \leftrightarrow p$
$[+\rho]\neg \phi \leftrightarrow \neg[+\rho]\phi$
$[+\rho](\phi \wedge \psi) \leftrightarrow [+\rho]\phi \wedge [+\rho]\psi$
$[+\rho]L\phi \leftrightarrow L[+\rho]\phi$
$[+\rho]I\phi \leftrightarrow I\phi$
$[+\rho]K\phi \leftrightarrow L[+\rho]\phi \wedge I\phi$
$[+\rho]A\sigma \leftrightarrow A\sigma$, for $\sigma \neq \rho$
$[+\rho]\hat{I}\phi \leftrightarrow \hat{I}\phi$
$[+\rho]A\rho \leftrightarrow \top$, for ρ sound. $[+\rho]A\rho \leftrightarrow \bot$, otherwise
$[+\rho]\langle RAD \rangle_\rho \phi \leftrightarrow \langle RAD \rangle_\rho [+\rho]\phi$

A Dynamic Epistemic Logic for Resource-Bounded Agents

$[-\rho](z_1 s_1 + \ldots + z_n s_n \geq z) \leftrightarrow (z_1 s_1 + \ldots + z_n s_n \geq z)$
$[-\rho]p \leftrightarrow p$
$[-\rho]\neg\phi \leftrightarrow \neg[-\rho]\phi$
$[-\rho](\phi \wedge \psi) \leftrightarrow [-\rho]\phi \wedge [-\rho]\psi$
$[-\rho]L\phi \leftrightarrow L[-\rho]\phi$
$[-\rho]I\phi \leftrightarrow I\phi$
$[-\rho]K\phi \leftrightarrow L[-\rho]\phi \wedge I\phi$
$[-\rho]A\sigma \leftrightarrow A\sigma$, for $\sigma \neq \rho$
$[-\rho]\hat{I}\phi \leftrightarrow \hat{I}\phi$
$[-\rho]A\rho \leftrightarrow \bot$
$[-\rho]\langle RAD \rangle_\rho \phi \leftrightarrow \langle RAD \rangle_\rho [-\rho]\phi$

Theorem 4 (Dynamic axiomatization) *The logic given by the system of Def. 11 and the reduction axioms of Th. 3 is sound and complete w.r.t.* **M**.

Proof. We first check that the reduction axioms are indeed valid in **M**. This, in combination with the result in Th. 2, suffices to prove the claim. □

4 Resource-bounded introspection

S5 models are reflexive, symmetric and transitive in order to capture properties of knowledge like factivity, positive and negative introspection. We have explained above that it is reasonable to impose reflexivity on our models too. However, due to the impossible worlds, symmetry and transitivity would not correspond to introspection. This is viewed as a desirable feature since avoiding unlimited introspection falls within the wider goal to model non-ideal agents. In analogy to our argumentation for resource-bounded factual reasoning, real agents are never fully introspective due to cognitive limitations therefore representation of reflective powers should be up to a certain modal depth (Ditmarsch & Labuschagne, 2007). In determining this depth, we adapt the tools used earlier for deductive reasoning.[9]

We first need to introduce introspective rules, add the respective terms and operators to the language (similar to the deductive case, e.g., $\langle \iota \rangle$ stands for "after applying introspective rule ι"), and fix cognitive costs. Then we need a model structure similar to the one provided by *Succession* to ensure that given sufficient resources, the agent can achieve higher and higher degrees of introspection. Extending V^* with the epistemic assertions satisfied

---

[9]Similar motivations fuel the works of Jago (2009) and Fervari and Velázquez-Quesada (2017), using actions of introspection.

at each world, we impose *Introspective Succession*: roughly, for every assertion composed by n K's there should be a successor world that validates an assertion with $(n+1)$ K's:[10]

Introspective Succession: For every $w \in W$, if: (i) $\psi \in V^*(w)$, where ψ is of the form $K^n\phi$ for some $\phi \in \mathcal{L}_P$ and $n = 0,\ldots,n$, and (ii) $\neg K\psi \notin V^*(w)$, then there is some $u \in W$ such that $V^*(u) = V^*(w) \cup \{K\psi\}$.

Negative introspective succession works similarly. For simplicity, let $pr(\iota)$ denote the initial assertion and $con(\iota)$ the new one. Then, defining the introspective radius and model transformation is analogous to the deductive case. In order to get w's introspective radius (w^ι), we replace ρ with ι in Def. 7, taking cases based on the new succession condition. Then the transformation becomes:

Definition 12 (Model transformation by applying an introspective rule) Let $M = \langle W^P, W^I, f, V, R, cp \rangle$ and $w \in W^P$. The transformation of the pointed model (M, w) by an application of ι is (M^ι, w), where:

- $W^\iota = W$, $V^\iota = V$, $R^\iota = R$, and $cp^\rho = cp - c(\iota)$.

- $f^\iota = g$ such that $g(v) = \begin{cases} \bigcup_{u \in f(w)} u^\iota, & \text{for } v = w \\ f(v), & \text{for } v \neq w \end{cases}$

This definition naturally leads to the truth conditions for applying ι: $M, w \models \langle \iota \rangle \phi$ iff $M, w \models cp \geq c_\iota$, $M, w \models A\iota$ and $M^\iota, w \models \phi$.

While the main idea (viewing both deduction and introspection as reasoning actions) is shared between this and earlier sections, the processes are discernible from a cognitive point of view. Both the costs and the availability of introspective rules may be treated differently, e.g., it might be that introspective rules are in principle always available and it is only the accumulated cost after each application that is responsible for limited reflection.

5 Conclusions

Overall, our dynamic logical framework overcomes logical omniscience but is also informed by empirical facts on the boundedness of real reasoners, as evinced by its formal results. The same considerations are extended to introspective abilities of agents. Meanwhile, we argue for the adequacy of this semantic approach against logical omniscience in terms of explanatory

---

[10] By making the model transitive, the condition is satisfied directly only for possible worlds.

power, but also show that it can be reduced to a syntactic one that enables the extraction of a sound and complete logic. Further work is especially needed on the optimal choice of cognitive parameters and on a multi-agent extension, building on the construction of inferential and introspective rules.

References

Baltag, A., & Renne, B. (2016). Dynamic epistemic logic. In E. N. Zalta (Ed.), *The Stanford Encyclopedia of Philosophy* (Winter 2016 ed.). Metaphysics Research Lab, Stanford University.

Berto, F. (2013). Impossible worlds. In E. N. Zalta (Ed.), *The Stanford Encyclopedia of Philosophy* (Winter 2013 ed.). Metaphysics Research Lab, Stanford University.

Berto, F., Smets, S., & Solaki, A. (2018). The logic of fast and slow thinking. *working paper*.

Bjerring, J. C., & Skipper, M. (2018). A dynamic solution to the problem of logical omniscience. *Journal of Philosophical Logic*.

Blackburn, P., de Rijke, M., & Venema, Y. (2001). *Modal Logic*. New York, NY, USA: Cambridge University Press.

Cherniak, C. (1986). *Minimal Rationality*. MIT Press.

Ditmarsch, H. V., & Labuschagne, W. (2007). My beliefs about your beliefs: A case study in theory of mind and epistemic logic. *Synthese, 155*(2), 191–209.

Fagin, R., & Halpern, J. Y. (1987). Belief, awareness, and limited reasoning. *Artif. Intell., 34*(1), 39–76.

Fagin, R., & Halpern, J. Y. (1994). Reasoning about knowledge and probability. *J. ACM, 41*(2), 340–367.

Fagin, R., Halpern, J. Y., Moses, Y., & Vardi, M. Y. (1995). *Reasoning About Knowledge*. MIT press.

Fervari, R., & Velázquez-Quesada, F. R. (2017). Dynamic epistemic logics of introspection. In *International Workshop on Dynamic Logic* (pp. 82–97).

Hintikka, J. (1962). *Knowledge and Belief: An Introduction to the Logic of the Two Notions*. Ithaca, N.Y.,Cornell University Press.

Hintikka, J. (1975). Impossible possible worlds vindicated. *Journal of Philosophical Logic, 4*(4), 475–484.

Jago, M. (2009). Epistemic logic for rule-based agents. *Journal of Logic, Language and Information, 18*(1), 131–158.

Johnson-Laird, P. N., Byrne, R. M., & Schaeken, W. (1992). Propositional reasoning by model. *Psychological Review*, *99*(3), 418–439.

Quesada, F. V. (2009). *Small steps in dynamics of information* (Vol. 1).

Rips, L. J. (1994). *The Psychology of Proof: Deductive Reasoning in Human Thinking*. Cambridge, MA, USA: MIT Press.

Smets, S., & Solaki, A. (2018). The effort of reasoning: Modelling the inference steps of boundedly rational agents. In L. S. Moss, R. de Queiroz, & M. Martinez (Eds.), *Logic, Language, Information, and Computation* (pp. 307–324). Berlin, Heidelberg: Springer Berlin Heidelberg.

Stanovich, K. E., & West, R. F. (2000). Individual differences in reasoning: Implications for the rationality debate? *Behavioral and Brain Sciences*, *23*(5), 645–665.

Stenning, K., & van Lambalgen, M. (2008). *Human Reasoning and Cognitive Science*. Boston, USA: MIT Press.

Stroll, A. (Ed.). (1967). *Epistemology*. New York: Harper and Rowe.

Tversky, A., & Kahneman, D. (1981, 02). The framing of decisions and the psychology of choice. *Science (New York, N.Y.)*, *211*, 453–458.

van Benthem, J. (2011). *Logical Dynamics of Information and Interaction*. Cambridge University Press.

van Ditmarsch, H., van der Hoek, W., & Kooi, B. (2007). *Dynamic Epistemic Logic* (1st ed.). Springer Publishing Company, Incorporated.

Verbrugge, R. (2009). Logic and social cognition. *Journal of Philosophical Logic*, *38*(6), 649–680.

Wansing, H. (1990). A general possible worlds framework for reasoning about knowledge and belief. *Studia Logica*, *49*(4), 523–539.

Wason, P. C. (1966). Reasoning. In B. Foss (Ed.), *New Horizons in Psychology* (pp. 135–151). Harmondsworth: Penguin Books.

Williamson, T. (2000). *Knowledge and its Limits*. Oxford University Press.

Anthia Solaki
University of Amsterdam
Institute for Logic, Language and Computation
The Netherlands
E-mail: `a.solaki2@uva.nl`